**Milestones in Drug Therapy**
**MDT**

# Aromatase Inhibitors

Edited by B.J.A. Furr

**Birkhäuser Verlag**
**Basel · Boston · Berlin**

**Editor**
Barrington J.A. Furr
Global Discovery
AstraZeneca
Mereside, Alderley Park
Macclesfield
Cheshire SK10 4TG
UK



**A CIP catalogue record for this book is available from the Library of Congress, Washington DC, USA**

**Bibliographic information published by Die Deutsche Bibliothek**
Die Deutsche Bibliothek lists this publication in the Deutsche Nationalbibliografie; detailed bibliographic data is available in the internet at http://dnb.ddb.de

ISBN 3-7643-7199-4 Birkhäuser Verlag, Basel - Boston - Berlin


Part of Springer Science+Business Media
Printed on acid-free paper produced from chlorine-free pulp. TFC ∞
Cover illustration: see p. 149. With the friendly permission of Evan Simpson

Printed in Germany
ISBN-10: 3-7643-7199-4
ISBN-13: 978-3-7643-7199-9
e-ISBN: 3-7643-7418-7

9 8 7 6 5 4 3 2 1

www. birkhauser.ch

# Contents

# List of contributors

Angela Brodie, Department Pharmacology & Experimental Therapeutics, University of Maryland, School of Medicine, Baltimore, MD 21201, USA; e-mail: abrodie@umaryland.edu

Aman Buzdar, Department of Breast Medical Oncology, The University of Texas M.D. Anderson Cancer Center, 1515 Holcombe Blvd 1354, Houston, TX 77030-4009, USA; e-mail: abuzdar@mdanderson.org

Colin D. Clyne, Prince Henry's Institute of Medical Research, P.O. Box 5152, Clayton VIC 3168, Australia; e-mail: colin.clyne@princehenrys.org

J. Michael Dixon, Edinburgh Breast Unit, Western General Hospital, Crewe Road, Edinburgh EH4 2XU, UK; e-mail: mike.dixon@ed.ac.uk

Barrington J.A. Furr, Research and Development, AstraZeneca, Mereside, Alderley Park, Macclesfield, Cheshire SK10 4TG, UK

Jürgen Geisler, Department of Medicine, Section of Oncology, Haukeland University Hospital, 5021 Bergen, Norway; e-mail: jurgen.geisler@helse-bergen.no

Anthony Howell, CRUK Department of Medical Oncology, Christie Hospital NHS Trust, Manchester, UK

Margaret E. Jones, Prince Henry's Institute of Medical Research, P.O. Box 5152, Clayton VIC 3168, Australia; e-mail: margaret.jones@princehenrys.org

Per Eystein Lønning, Department of Medicine, Section of Oncology, Haukeland University Hospital, 5021 Bergen, Norway; e-mail: per.lonning@helse-bergen.no.

William R. Miller, Breast Unit, Paderewski Building, Western General Hospital, Edinburgh, EH4 2XU, UK; e-mail: w.r.miller@ed.ac.uk

Robert J. Paridaens, University Hospital Gasthuisberg, Katholieke Universiteit Leuven, Herestraat 49, 3000 Leuven, Belgium; e-mail: robert.paridaens@uz.kuleuven.be

Evan R. Simpson, Prince Henry's Institute of Medical Research, P.O. Box 5152, Clayton VIC 3168, Australia; e-mail: evan.simpson@phimr.monash.edu.au

Alan Wakeling, Department of Cancer and Infection Research, AstraZeneca Pharmaceuticals, Macclesfield, UK; e-mail: Alan.Wakeling@astrazeneca.com

# Preface

It is over 100 hundred years since the Glaswegian surgeon James Beatson showed that many breast cancers were dependent on the ovaries for their growth. Some time later oestrogen was shown to be the ovarian factor responsible for the development and growth of many breast cancers in both premenopausal and postmenopausal women, in whom it was produced from adrenal androgens by peripheral tissues and by the tumours themselves. As a consequence, endocrine therapies for breast cancer have been developed that lead to either a reduction in oestrogen production or antagonism of its action.

In premenopausal women surgical removal of the ovaries or ablation by radiation have largely been superseded by therapy with gonadotrophin-releasing hormones, like Zoladex, that produce an effective medical oophorectomy. In postmenopausal women inhibition of the enzyme aromatase, which catalyses the last step in oestrogen biosynthesis, has long been a target for the pharmaceutical industry. The first aromatase inhibitor to be introduced, aminoglutethimide, proved effective but was tarnished by a lack of selectivity. It also caused loss of production of adrenal corticosteroid hormones and so had to be given with cortisone replacement. The associated toxicity gave an opportunity for the oestrogen receptor antagonist, tamoxifen, which was much better tolerated, to become established as the primary endocrine treatment for advanced and early breast cancer and as an adjuvant to surgery.

Second-generation aromatase inhibitors were developed that had greater selectivity but poor bioavailability and so their use was restricted. The advent of the third-generation aromatase inhibitors – anastrozole, letrozole and exemestane – provided far more potent, selective and orally active therapies that could be given once daily and these are now challenging the dominance of tamoxifen at all stages of breast cancer treatment. Indeed, it is likely that they will supplant tamoxifen because of their improved efficacy and tolerability.

Chapters in this volume outline the history and basic biochemistry of aromatase inhibitors, their efficacy in disease models and clinical pharmacology. In view of the extensive experience with these third-generation compounds individual chapters on anastrozole, letrozole and exemestane have been written by clinicians well versed in their use. An overview chapter looks objectively at the field and draws general conclusions about the value of these inhibitors in the treatment of breast cancer and the strength of the clinical data that underpins their use. The careful study of aromatase and oestrogen receptor-knockout mice has elucidated several novel and subtle actions that may have important bearing, both on the long-term use of aromatase inhibitors in breast cancer and on other uses to which they might be put. The chapter on this topic beautifully complements both the preclinical and clinical reviews.

The additional potential uses of aromatase inhibitors outside of breast cancer have been reviewed in the final chapter.

It has been my privilege to work with the outstanding preclinical and clinical scientists who have made major contributions to the development of aromatase inhibitors and an understanding of the role of the aromatase in pathobiology.

Barrington J.A. Furr October 2005

Aromatase Inhibitors
Edited by B.J.A. Furr



# Background and development of aromatase inhibitors

William R. Miller

*Breast Unit, Paderewski Building, Western General Hospital, Edinburgh EH4 2XU, UK*

## Introduction

The natural history of breast cancer suggests that many tumours are dependent upon oestrogen for their development and continued growth [1]. As a consequence it might be expected that oestrogen deprivation will both prevent the appearance of these cancers and cause regression of established tumours [2]. This provides the rationale behind hormone prevention of breast cancer and endocrine management of the disease. Over the last 25 years hormone therapy has progressed from the irreversible destruction of endocrine glands, as achieved by either surgery or radiation (with high co-morbidity), to the use of drugs that reversibly suppress oestrogen synthesis or action (with minimal side effects). In terms of inhibiting oestrogen biosynthesis, it is relevant that primary sites of oestrogen production differ according to menopausal status. Thus in premenopausal women the ovaries are the major source of oestrogen whereas peripheral tissues such as fat, muscle and the tumour itself are more important in postmenopausal patients [3]. In using drugs to block biosynthesis, it is most attractive to employ agents which specifically affect oestrogen production irrespective of site. Mechanistically, this is most readily achieved by inhibiting the final step in the pathway of oestrogen biosynthesis, the reaction which transforms androgens into oestrogens by creating an aromatic ring in the steroid molecule (hence the trivial name of aromatase for the enzyme catalysing this reaction).

Although the first aromatase inhibitors to be used therapeutically could be shown to produce drug-induced inhibition of the enzyme and therapeutic benefits in patients with breast cancer [4], they were not particularly potent and lacked specificity, which often produced side effects unrelated to oestrogen deprivation. However, subsequently, second-generation drugs were developed [5] and most recently third-generation inhibitors have evolved which possess remarkable specificity and potency. Initial results from clinical trials suggest these agents will become the cornerstones of future endocrine therapy. The evolution of aromatase inhibitors is a classic example of successful rationale drug development and is the subject of this review.

## Aromatase

Oestrogens are the end-products of a sequence of steroid transformations (Fig. 1). Blockade of any conversion in the pathway potentially leads to decreased oestrogen production, but more specific suppression will result from inhibition of the final step that is unique to oestrogen biosynthesis. This reaction that changes androgens into oestrogens is complex. It involves 3-hydroxylations, each using NADPH as an electron donor [6], to eliminate the C-19 methyl group and render the steroid A ring aromatic (Fig. 2). A single enzyme is responsible [7], which possesses a prosthetic specific cytochrome P450 (P450 arom) and a ubiquitous flavoprotein NADPH cytochrome P450 reductase [8]. The key role of aromatase in oestrogen biosynthesis has generated enormous interest in putative inhibitors of the enzyme and their use as therapy against endocrine responsive tumours.

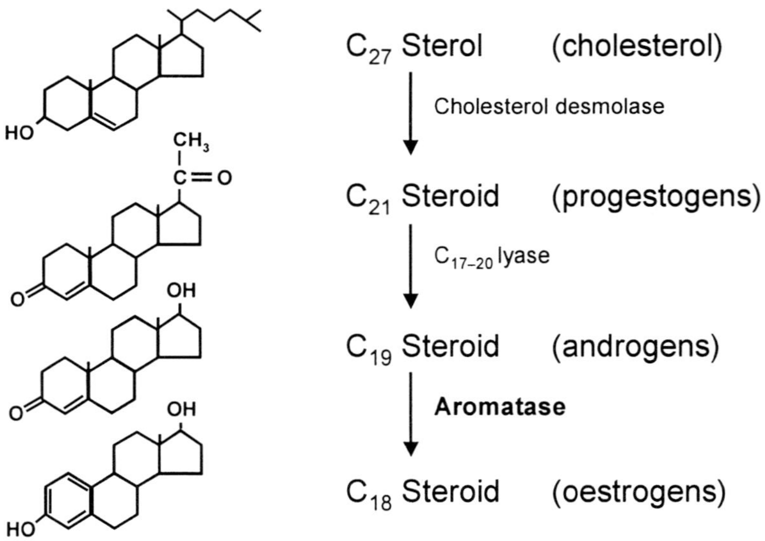


Figure 1. Classical pathway of oestrogen biosynthesis from cholesterol.

## Aromatase inhibitors

Inhibitors of aromatase have been subdivided into two main groups according to their mechanism of action and structure (Fig. 3). Type I inhibitors associate with the substrate-binding site of the enzyme and invariably have an androgen structure (and are often referred to as steroidal inhibitors). In contrast, type II inhibitors interact with the cytochrome P450 moiety of the system and, structurally, the majority are azoles (Fig. 3) and 'non-steroidal'.

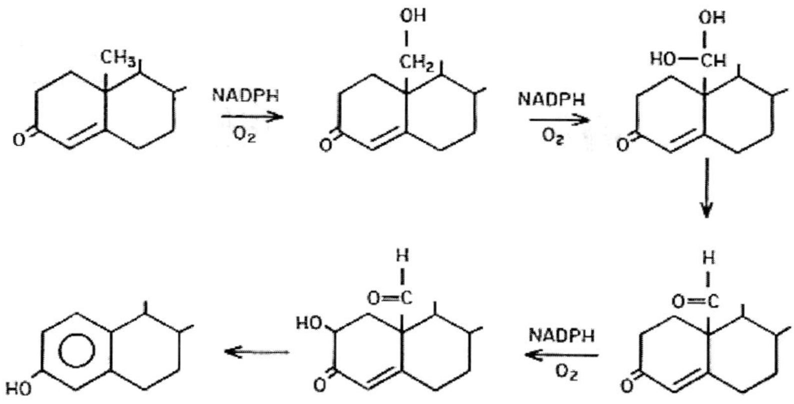


Figure 2. Proposed mechanism of oestrogen biosynthesis.

Type I agents are generally more specific inhibitors than type II. Some type I inhibitors, such as formestane and exemestane, have negligible inhibitory activity *per se* but, on binding to the catalytic site of the enzyme, are metabolized into intermediates which attach irreversibly to the active site of the enzyme, thus blocking activity [9]. These agents have been termed suicide inhibitors since the enzyme becomes inactivated only as a consequence of its own mechanism of action. Such mechanism-based inhibitors are particularly

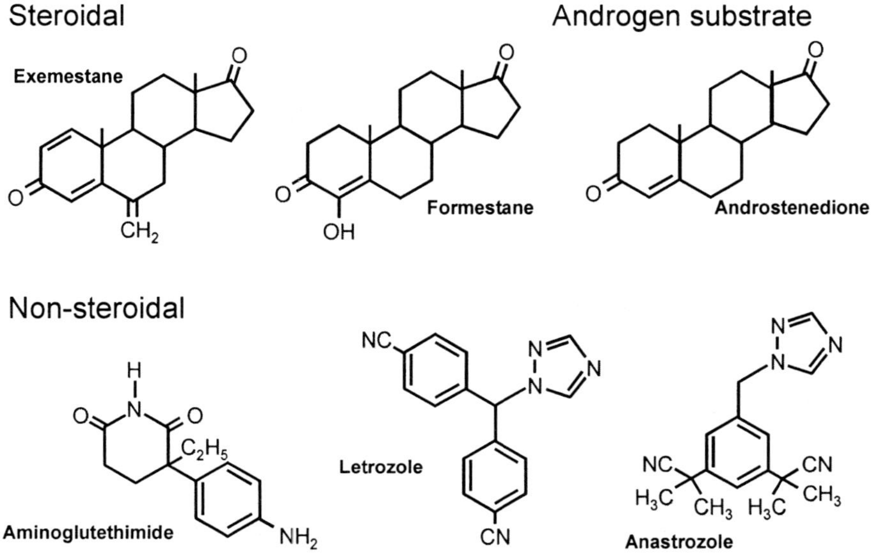


Figure 3. Different classes of aromatase inhibitor. Steroidal inhibitors are androgen analogues and non-steroidal inhibitors, such as aminoglutethimide, letrozole and anastrozole, are azoles.

specific as they inactivate only the enzyme for which they are metabolic substrates. Prolonged effects may occur *in vivo* because the enzyme is inactivated even after the drug is cleared from the circulation. Resumption of oestrogen production depends on the synthesis of new aromatase molecules.

The properties of type I inhibitors are to be contrasted with type II agents, which do not destroy the enzyme and whose actions are usually reversible and dependent upon the continued presence of inhibitor (see below). Type II inhibitors interact with the haem group of the cytochrome P450 moiety within the aromatase enzyme [10]. They may lack specificity because other enzymes, including other steroid hydroxylases, also have cytochrome P450 prosthetic groups and may therefore be inhibited [11]. Specificity of this binding is determined by fit into the substrate-binding site of aromatase as opposed to that of other cytochrome P450 enzymes. Because the amino acid sequence of P450 arom is distinct from other members of the P450 cytochrome family [12], it has been possible to develop drugs with selectivity towards the cytochrome P450 in aromatase, permitting more specific inhibition [11].

The evolution of aromatase inhibitors has seen the development of agents of both classes that have progressively increased in both specificity and potency with each new generation (Tab. 1).

Table 1. Classification of aromatase inhibitors

| | Inhibitor | | |
|---|---|---|---|
| Generation… | First | Second | Third |
| Type I (steroidal) | Testololactone | Formestane | Exemestane |
| Type II (non-steroidal) | Aminoglutethimide | Fadrozole | Anastrozole<br>Letrozole |

## First-generation drugs, the prototype aromatase inhibitors

It is only in relatively recent years that clinical trials have employed drugs designed specifically as aromatase inhibitors. Early inhibitors, such as testololactone and aminoglutethimide, were used without the knowledge that they had anti-aromatase properties [13–16]. For example, testololactone was given as an androgen [17] and aminoglutethimide was introduced as a form of medical adrenalectomy [14, 15, 18].

The development of aminoglutethimide as an endocrine therapy for breast cancer is particularly informative and worthy of further consideration. Thus aminoglutethimide first entered preliminary trials in advanced breast cancer as a result of the observation that it inhibited adrenal steroidogenesis during its earlier investigation as an antiepileptic [19]. The basis of the use of aminoglutethimide in this context was that adrenal androgens form the principal substrate for the synthesis of plasma oestrogens by aromatase in the peripheral tissues of postmenopausal women: removal of these androgens would therefore be expected to elicit the attenuation of the oestrogenic stimulus to the breast carcinoma by a process termed medical adrenalectomy [14]. The drug was given in sufficient doses to inhibit the production of adrenal steroids, and replacement corticoids were needed to avoid potential problems of adrenal insufficiency. Subsequently (during the early 1970s), Thompson and Siiteri [20] established that aminoglutethimide was an inhibitor of the aromatase enzyme, and a classic paper by Santen and colleagues [21] demonstrated that the aminoglutethimide-corticoid regimen blocked peripheral conversion of androgens to oestrogen and suppressed circulating oestrogens in postmenopausal women with breast cancer. This led to the development of the concept of a dual mode of action for aminoglutethimide in which the drug both suppressed adrenal androgen synthesis and inhibited the conversion of any residual androgen to oestrogen. However, debate continued as to whether the anti-tumour action of aminoglutethimide regimes primarily resulted from effects on adrenal steroidogenesis or from those on peripheral aromatase. Evidence that the latter were more important derived from experimentation using low doses of aminoglutethimide that could be given in the absence of corticoid replacement [22]. The aromatase system is about 10-fold more sensitive to aminoglutethimide than cholesterol side-chain cleavage [23]. Low-dose regimes of aminoglutethimide-hydrocortisone were more selective against aromatase [24] but they still elicited anti-tumour responses [25]. These remissions produced by aminoglutethimide in the absence of corticoid replacement [22, 26] substantiate the hypothesis that the aminoglutethimide component of the conventional regime was responsible for anti-tumour effects.

The response rate, duration and site of response to the standard daily dosage regime of aminoglutethimide (250 mg, four times daily) plus hydrocortisone (20 mg, twice daily) in postmenopausal women with advanced breast cancer were similar to those reported for other endocrine therapies [27–31]. In four large series of unselected patients response rates varied from 28 to 37%, with an average value of 33%, with about a further 15% of patients benefiting from disease stabilization. Patients with a previous objective response to hormone therapy were twice as likely to respond than those who had failed endocrine treatment [27]. Median duration of response to aminoglutethimide was about 14 months [27, 32]. In general, soft tissue and lymph nodes responded better than visceral sites [33].

The presence of oestrogen receptor (ER) in tumours predicts for response to aminoglutethimide [34, 35]. Thus response rates in ER-negative tumours are usually less than 10%, whereas those in ER-positive tumours can exceed 50% [33]. This would substantiate the idea that the major effects of aminoglutethimide are mediated by oestrogen deprivation and would explain why the drug is less successful in premenopausal women, in whom the drug does not effectively reduce oestrogen levels [36].

Aminoglutethimide is effective as a second-line endocrine therapy and almost one-half of patients responding to tamoxifen, adrenalectomy or hypophysectomy may have a further response to aminoglutethimide given subsequently [33]. The drug may decrease oestrogens in both adrenalectomized and hypophysectomized patients [37].

The interrelationship between response to aminoglutethimide and tamoxifen is particularly interesting. Whereas aminoglutethimide is effective in about 30% of patients after tamoxifen (20% non-responders and 60% responders to tamoxifen), the anti-oestrogen less frequently causes remission after aminoglutethimide [38–40]. Furthermore, the combination of tamoxifen and aminoglutethimide is not significantly more successful than the two drugs given singly or sequentially [41, 42]. The greater tolerability problems with aminoglutethimide plus corticoids [43] and the lesser side effects of tamoxifen also suggest that the optimal sequence of treatment is tamoxifen before aminoglutethimide.

Although this early work was important in establishing that aromatase inhibition with aminoglutethimide was a viable method of treating postmenopausal patients with advanced breast cancer, it was clear that aminoglutethimide was far from an ideal agent. The drug was only partially effective in suppressing plasma oestrogen levels, and its lack of specificity required the routine use of glucocorticoid replacement. The lack of specificity of aminoglutethimide largely results from its actions on other cytochrome P450 systems [11]. Most significantly, aminoglutethimide had several marked side effects, including lethargy and somnolence extending to ataxia as well as nausea and vomiting [19]. Thus the scene was set for the pharmaceutical industry to derive more specific, fully effective and better-tolerated aromatase inhibitors.

## Second-generation drugs

Among the next generation of aromatase inhibitors to reach the clinic, the most notable were the steroidal drug, formestane (4-hydroxyandrostenedione (4-OHA)), and the non-steroidal imadazole, fadrozole (CGS16949A).

4-OHA was one of about 200 compounds which were specifically designed and screened as aromatase inhibitors by Drs Harry and Angela Brodie in the 1970s [44, 45]. It bound competitively with androgen substrate but, in addition, appeared to be converted by the aromatase enzyme to reactive intermediates that bound irreversibly to the enzyme and produced a time-dependent inactivation of aromatase activity [44, 46]. 4-OHA was about 60-fold more potent than aminoglutethimide in inhibiting aromatase activity in placental microsomes [9]. The agent caused regression of hormone-dependent mammary tumours in experimental animals [44, 45] and chronically abolished peripheral aromatase in rhesus monkeys [46].

Pharmacological and endocrinological studies in postmenopausal women confirmed efficacy but, when given orally, 4-OHA had poor biological activity as measured by both inhibition of aromatization *in vivo* [47–49] and sustained oestrogen suppression [50]. This resulted from the glucuronidation of the critical 4-hydroxy group through first-pass liver metabolism. Further studies and clinical use focused on the intramuscular administration of the drug.

Intramuscular administration of 250 mg every second week was the preferred schedule, inhibiting peripheral aromatase inhibition by 85% and suppressing circulating oestradiol by about 65% [51]. A small recovery of plasma oestrogens occurred prior to the next injection [48, 52], but nonetheless the regime was chosen for routine clinical use because of greater tolerability problems with higher doses [53]. Objective tumour regressions were observed in 23–39% of patients and disease stabilization in a further 14–29%. As with aminoglutethimide, patients who had a previous response to other hormone therapy were much more likely to respond to 4-OHA. Interestingly, three of 14 patients previously treated with aminoglutethimide subsequently responded to 4-OHA, suggesting that a more potent aromatase inhibitor may produce further remission after the benefits of a less powerful inhibitor have been exhausted. Several phase II studies confirmed the clinical efficacy of 4-OHA [53]. In one phase III study comparing formestane to tamoxifen as first-line treatment of advanced breast cancer, no difference in response rate or survival was recorded, but the median duration of response was significantly longer for tamoxifen [54]. Another phase III study compared formestane as second-line treatment to megesterol acetate and found no difference in response rate, time to progression, or survival [55]. The particular advantages of 4-OHA were its low toxicity, its specificity and the lack of need for corticoid replacement.

Second-generation type II inhibitors were also developed with greater selectivity and potency than their first generation counterparts. For example, fadrozole is an imidazole derivative of aminoglutethimide which inhibited the aromatase system in human placenta and rodent ovary with about 400–1000-fold greater potency than aminoglutethimide [56]. At concentrations that maximally inhibit aromatase, unlike aminoglutethimide, the drug had relatively small effects on other cytochrome P450-related enzymes [56]. This meant the drug could be administered to patients without the need for corticoid replacement.

Animal studies showed that fadrozole was an effective anti-tumour agent. For example, the drug produces marked regression of dimethyl-benzanthracene (DMBA)-induced mammary carcinomas [57].

A daily dose (2 mg) of fadrozole produced comparable aromatase suppression (as measured by urinary and plasma oestrogens) as the standard regime of aminoglutethimide (1000 mg plus 40 mg of hydrocortisone) [58]. Two further studies using a dose of 2 mg/day reported tumour remissions in heavily pretreated postmenopausal women with advanced breast cancer: in one investigation five of 31 patients experienced a partial or complete response [59], and in the other two of 15 patients had a partial response and a further seven patients had stabilization of disease [60]. Side effects from fadrozole were few and the drug was given orally. These results are in keeping with (i) a further study of 80 previously treated postmenopausal women with advanced breast cancer who were randomized to receive 1 or 4 mg of fadrozole per day, complete responses being documented in 10% and partial responses in 13% of patients, with no significant differences between doses [61], and (ii) a double-blind randomized multicentre study using doses of 1, 2 and 4 mg/day which observed objective responses in 16% of 350 women who had already received tamoxifen either for treatment of advanced cancer or as an adjuvant for early disease [62]. A similar response rate has been reported in recurrent breast cancer after tamoxifen failure [63]. Fadrozole was also as effective as megestrol acetate in postmenopausal women progressing after anti-oestrogen treatment [64]. A phase III comparative trial of fadrozole (2 mg) *versus* tamoxifen (20 mg) as first-line treatment for postmenopausal advanced breast cancer [65] reported objective responses in 16% of fadrozole-treated patients compared with 24% of tamoxifen patients (another 50% of women in each group also experienced disease stabilization), the difference between the groups not reaching statistical significance.

Whereas fadrozole is a highly potent compound, it has a relatively short half-life, which accounts for its poorer *in vivo* activity compared with triazole inhibitors that are cleared more slowly [66]. Doubts have also been raised about the specificity of fadrozole since it can also suppress cortisol and aldosterone synthesis [67, 68], although these effects may not be of clinical significance [69]. At present, this compound is used widely only in Japan.

## Third-generation inhibitors

These aromatase inhibitors include anastrozole [70], letrozole [71, 72] and exemestane (vorozole was withdrawn early in development despite being highly potent and specific [73, 74]). Both letrozole and anastrozole are triazoles which have a flat aromatic ring providing a good fit with the substrate-binding site of the enzyme. Additionally, there is a moiety within the ring structure that coordinates with the aromatase haem iron and effectively inhibits the hydroxylation reactions necessary for aromatization. The combination of haem-group-binding and active-site binding provide high potency and greater target specificity. Exemestane is an androgen analogue that inactivates aromatase in the same manner as formestane.

Anastrozole, letrozole and exemstane are all substantially more potent than aminoglutethimide in terms of inhibiting *in vitro* aromatase activity (Tab. 2). Whereas the drug concentrations required are micromolar for aminoglutethimide, those for letrozole, anastrozole and exemestane are nanomolar. The superior pharmacokinetic profiles of third-generation drugs also mean they are even more effective *in vivo*. In this respect, milligram daily doses of anastrozole, letrozole and exemestane effectively inhibit whole-body aromatization (Tab. 3), and circulating oestrogens may fall below detectable levels [75]. It is thus worth considering each of these drugs in further detail.

Table 2. Inhibition of aromatase activity in whole-cell and disrupted-cell preparation

| | Placental microsomes | | Breast cancer homogenates | | Mammary fibroblast cultures | |
|---|---|---|---|---|---|---|
| | $IC_{50}$ (nM) | Relative potency | $IC_{50}$ (nM) | Relative potency | $IC_{50}$ (nM) | Relative potency |
| Aminoglutethimide | 3000 | 1 | 4500 | 1 | 8000 | 1 |
| Anastrozole | 12 | 250 | 10 | 450 | 14 | 570 |
| Letrozole | 12 | 250 | 2.5 | 1800 | 0.8 | 10 000 |
| Formestane | 50 | 60 | 30 | 150 | 45 | 180 |
| Exemestane | 50 | 60 | 15 | 300 | 5 | 1600 |

Table 3. Aromatase inhibition *in vivo*. Data from [75, 133]. Drugs given orally except for formestane, which was given intramuscularly (i.m.).

| | Inhibition (%) | Residual activity (%) |
|---|---|---|
| Exemestane | 97.9 | 2.1 |
| Formestane (i.m.) | 91.9 | 8.1 |
| Aminoglutethimide | 90.6 | 9.4 |
| Anastrozole | 96.7 | 3.3 |
| Letrozole | 98.9 | 1.1 |

## *Anastrozole*

This triazole is a potent aromatase inhibitor *in vivo*, with daily doses of 1 and 10 mg given to postmenopausal women showing a mean aromatase suppression of 96.7 and 98.1% respectively. Plasma oestrone, oestradiol and oestrone sulphate are reduced by at least 80%, with many treated patients having levels of oestrone and oestradiol beneath the level of sensitivity of the assays. This occurs without detectable changes in other steroid hormones [76]. Impressive anti-tumour effects have also been observed in patients with breast cancer but these are detailed in other chapters.

## *Letrozole*

Letrozole potently inhibits peripheral aromatase and suppresses endogenous oestrogens in postmenopausal women. At 0.5 and 2.5 mg/day, letrozole inhibits peripheral aromatase by >98% [77]. Doses as low as 0.1 mg/day can suppress circulating levels of oestrone, oestrone sulphate and oestradiol by more than 95% within 2 weeks of treatment [78], these effects being greater than those observed after the use of second-generation inhibitors. In a direct comparison between letrozole and the second-generation inhibitor fadrozole, letrozole was more effective, suppressing plasma oestrogen concentrations to undetectable levels (>95% baseline) at all doses investigated (0.1–5 mg/day) while fadrozole (2–4 mg daily) only achieved above 70% suppression [78]. No substantial suppression of cortisol and aldosterone levels is evident even at doses of 5 mg/day (and *in vitro* aldosterone production is only inhibited with 10 000-fold higher concentrations than those required to inhibit oestrogen synthesis [79]). Recently results from a randomized cross-over study of letrozole and anastrozole have been published [80]. Treatment with letrozole suppressed levels of *in vivo* aromatization below the detection limit of the assays (>99.1% inhibition) in all 12 patients. In contrast, anastrozole treatment produced this degree of suppression inhibition in only one of 12 cases. The mean inhibition of aromatization (97.3% for anastrozole *versus* >99.1% for letrozole) was significantly different ($P = 0.0022$). This corresponded to a 10-fold lower residual level of aromatization during letrozole treatment compared to anastrozole (0.006 *versus* 0.059%). It still remains to be determined whether these differences in suppression of aromatase translate into differences in clinical benefit.

Clinically, letrozole produces tumour remission in postmenopausal women with breast cancer resistant to other endocrine treatments and chemotherapy and these are described in other chapters. However, it is important to note that letrozole had greater efficacy than the first-generation inhibitor aminoglutethimide in terms of time to progression ($P = 0.008$) and overall survival ($P = 0.002$; median, 28 *versus* 20 months) [81]. This last comparison emphasizes the improvement in efficacy that has occurred by virtue of the development of the new non-steroidal aromatase inhibitors and also emphasizes the improvement in tolerability: adverse events were 29% with letrozole *versus* 46% with aminoglutethimide.

### *Exemestane*

Exemestane is an orally active steroidal inhibitor. A dose of 25 mg/day results in an inhibition of aromatase *in vivo* by 98%. Exemestane will reduce oestrogen levels in patients relapsing on the first-generation inhibitor aminoglutethimide [82].

## Advantages/disadvantages of aromatase inhibitors as endocrine therapy for breast cancer

Specific inhibitors of the aromatase system have several advantages over more general endocrine therapies such as surgical ablation of endocrine glands. First, the actions of aromatase inhibitors are not totally irreversible and, should therapy prove ineffective, oestrogen levels usually return to normal on discontinuation of treatment [83]. Second, a 'pure' aromatase inhibitor will specifically decrease oestrogen alone whereas ablation of endocrine organs additionally affects other steroid hormones. As a consequence, aromatase inhibitors are associated with fewer side effects and lower morbidity. Third, aromatase inhibitors have the potential for total blockade of oestrogen production since biosynthesis is not restricted to classical endocrine glands but occurs at multiple peripheral sites including the majority of breast cancers [84]. Because the aromatase complex appears similar in both endocrine and peripheral tissue [85], inhibitors are capable of suppressing oestrogen levels beyond those achievable by surgical ablation of endocrine glands [86].

Conversely, specific aromatase inhibitors have theoretical disadvantages in treating oestrogen-dependent breast cancers in that they will not affect exogenously derived oestrogen or levels of other types of steroids such as androstenediol, which may be oestrogenic [87]. In addition, they are unproven as effective therapy in premenopausal women [36, 88]. Earlier inhibitors such as aminoglutethimide were largely ineffective at reducing circulating oestrogens and did not produce clinical benefit [36, 88, 89]. It appears that the high levels of aromatase in the ovary and compensatory hypothalamic/pituitary feedback loops were obstacles to inhibition of ovarian oestrogen production [4, 89] (they may also cause ovarian hyperplasia and cysts). Whether the later generation of aromatase inhibitors will be more successful in this setting is still to be determined. Currently, aromatase inhibitors are used in combination with agents which block the compensatory feedback loops and render premenopausal women postmenopausal. The most promising regime is an aromatase inhibitor in combination with a luteinizing hormone-releasing hormone (LHRH) agonist [90].

## Differences between anti-oestrogens and aromatase inhibitors

It is important to note that advantages of reversibility and specificity, irrespective of oestrogen source, are shared by aromatase inhibitors and anti-oestrogens (selective oestrogen receptor modulators; SERMs). However, the mechanisms of action of SERMs and aromatase inhibitors are sufficiently different that tumour response to the two agents is not mutually exclusive, even though both reduce oestrogen signalling within breast cancers. Different effects on endogenous oestrogens and interactions with the ER may be particularly important. In terms of the former, aromatase inhibitors reduce endogenously synthesized oestrogens whereas SERMs such as tamoxifen do not inhibit synthesis and oestrogen levels remained unaltered [91] (or, in the case of premenopausal women, may increase [92, 93]). This difference may be critical in certain circumstances because oestrogen metabolites may act independently of ER-mediated mechanisms [94]. Since these processes may include genotoxic damage there might be additional advantages in using aromatase inhibitors to prevent cancer. Conversely, whereas specific aromatase inhibitors reduce levels of oestrogen synthesized endogenously, they will not block the activity of exogenous oestrogens or oestrogen mimics such as polychlorinated biphenyls (PCBs), nonyl phenols, phyto-oestrogens and certain androgens, which may interact with the ER [87, 95–97]. In contrast, tamoxifen will interfere with ER signalling irrespective of ligand. However, given that third-generation aromatase inhibitors appear more effective as anti-tumour agents than tamoxifen [98–103], it may be that oestrogen mimics are generally less influential than classical oestrogens in the natural history of breast cancers [104].

A further difference between aromatase inhibitors and the most widely used anti-oestrogen, tamoxifen, is that specific aromatase inhibitors do not interact directly with the ER and are without oestrogen agonist activity, whereas tamoxifen binds directly to the ER. This can most readily be illustrated by the effects of treatment on the expression of a classical marker of oestrogenic activity, the progesterone receptor. Thus, whereas aromatase inhibitors reduce the tumour expression frequently to zero, a common effect of tamoxifen is to increase expression [105]. The general phenotype of an aromatase inhibitor-treated tumour is ER-positive/progesterone receptor-negative, whereas that of a tamoxifen-treated tumour is ER-poor/progesterone receptor-rich. This may have implications for the sequence in which the agents are used during treatment. Because of these differences between tamoxifen and specific aromatase inhibitors, it might be expected that aromatase inhibitors will (i) be effective in tamoxifen-resistant tumours, (ii) produce increased response rates (if oestrogen suppression is more effective than oestrogen antagonism), (iii) produce responses more quickly than tamoxifen (aromatase inhibitors reduce oestrogen levels rapidly [72, 106], whereas the concentrations of tamoxifen for effective oestrogen blockade accumulate relatively slowly [107]) and (iv) be less effective in the presence of tamoxifen (if tamoxifen is more likely to have agonist properties in the low-oestrogen environment induced by aromatase inhibitors).

## Response and resistance to aromatase inhibitors

Whereas increasing numbers of patients with breast cancer derive benefit from aromatase inhibitors, as with other forms of endocrine therapy, many tumours do not respond. Even in responding patients, remission is not generally permanent and disease may recur. It is thus important to identify markers that are associated with response and mechanisms by which resistance occurs.

The best single marker for predicting response is tumour ER status; responses are usually associated with ER positivity and receptor-negative tumours rarely respond [1, 33, 35, 108]. However, the presence of ER does not guarantee a successful outcome to treatment, and response rates may be as low as 40–50% in ER-positive tumours. There is thus a need to find other predictive indices. Interestingly, overexpression of the cerbB signalling receptors, associated with resistance to tamoxifen, does not appear to reduce response rates to third-generation aromatase inhibitors [109, 110].

Since aromatase inhibitors achieve their benefit by causing oestrogen deprivation, many of the mechanisms by which resistance occurs are likely to be shared by other forms of endocrine deprivation. These include the loss of ERs with treatment (although this seems to occur only rarely) [111–113], the presence of defective ERs or oestrogen signalling [114, 115], the outgrowth of hormone-insensitive cells [116], ineffective oestrogen suppression and/or endocrine compensation [117, 118], and a switch to dependence on other mitogens [119, 120].

There may also be mechanisms specific to aromatase inhibitors [113]. Reference has already been made to premenopausal women in whom high ovarian aromatase is difficult to block. Although aromatase activities in peripheral sites in postmenopausal women are lower than in the premenopausal woman's ovary, levels may be elevated under certain conditions. For example, aminoglutethimide-hydrocortisone may paradoxically induce aromatase activity in breast cancer [121]. This could potentially reduce the efficacy of aminoglutethimide in patients on prolonged therapy, and may account for the beneficial effects which have been reported for the use of more potent aromatase inhibitors in aminoglutethimide-treated patients.

It is also possible that mutant/abnormal forms of the aromatase enzyme may be resistant to certain aromatase inhibitors. Interestingly, therefore, studies in which site mutations are introduced into the cDNA encoding for aromatase [122] have generated a phenotype displaying resistance to 4-OHA (without changing sensitivity to aminoglutethimide or affecting aromatase activity). These characteristics are also observed in certain primary breast cancers [123, 124], although molecular analysis has failed to provide evidence of a mutation in the aromatase gene [125]. Irrespective of the cause of the phenotype, certain tumours may be more sensitive/resistant to individual aromatase inhibitors. Additionally, since steroidal and non-steroidal aromatase inhibitors have a different mechanism of action, non-cross resistance can occur and has been reported in the clinical setting [126, 127].

## Future expectations and concluding perspectives

Third-generation aromatase inhibitors appear (i) to be extremely potent and highly specific inhibitors of the aromatase enzyme and able to suppress *in vivo* peripheral aromatase and circulating levels of oestrogens in postmenopausal women beyond the effects of previous inhibitors, (ii) to have antitumour effects in postmenopausal women with breast cancer which are at least as beneficial as other established endocrine agents and (iii) to be remarkably well tolerated, having no greater side effects than might be expected from oestrogen suppression. The expectation is, therefore, that they will have greater utility than other aromatase inhibitors not only in terms of increased response rates and more enduring responses in patients with breast cancer but a wider application in women without breast cancer with regard to cancer prevention and treatment of benign conditions.

With regard to increased duration and incidence of response, if breast cancers are composed of cellular clones with different oestrogen sensitivity, relapse might occur as a consequence of the outgrowth of cells that can exist on minimal hormone levels. Agents that produce greater oestrogen suppression might, therefore, be expected to prevent the outgrowth of such clones and thereby to extend duration of response. Similarly, some tumours that do not respond to endocrine therapy may not be totally insensitive to hormones but require only small amounts of oestrogen. More potent endocrine agents could, therefore, be effective in these cases. In this respect, third-generation inhibitors may cause remissions in tumours that are insensitive to other aromatase inhibitors and endocrine agents. Clinical evidence pertinent to these concepts is reviewed in other chapters.

Because aromatase inhibitors attenuate oestrogen action by reducing concentration of oestrogens, they may have additional benefits associated with non-ER mediated effects. In this respect it is clear that the oestrogen molecule may have pleiotropic effects, not all of which are transduced through ER. It has, therefore, been argued that aromatase inhibitors may have a particular role in the prevention of cancer and the treatment of certain benign conditions [128–132].

Questions relating to which aromatase inhibitor to use in which setting still need to be answered. Third-generation inhibitors share similar profiles in terms of potency, specificity, clinical efficacy and tolerability but there are differences in pharmacology, structure and mode of action. To determine whether these differences will impact on clinical benefit requires results from direct trial comparisons and these data are not substantially available. There is also the issue of whether even more potent inhibitors should be developed. Given that current third-generation inhibitors are already extremely specific and potent and that the efficacy and toxicity profiles of long-term use have not been fully evaluated, it seems premature to search for even more powerful drugs.

The final perspective is that the use of inhibitors that produce complete and specific blockade of oestrogen biosynthesis offers the opportunity to learn more about the role of that system in health and disease. There is therefore no doubting that observations derived from therapeutic interventions and laboratory experiments with the third-generation aromatase inhibitors will provide fundamental knowledge about the role of aromatase and oestrogen in hormone-dependent processes.

## References

1 Miller WR (1996) Aromatase inhibitors. *Endocr Relat Cancer* 3: 65–79
2 Henderson IC, Canellos GP (1980) Cancer of the breast: the past decade. *New Engl J Med* 302: 17–30: 78–90

3 Miller WR (1990) Endocrine treatment for breast cancers: biological rationale and current progress. *J Steroid Biochem Mol Biol* 37: 467–480

4 Miller WR (1989) Aromatase inhibitors in the treatment of advanced breast cancer. *Cancer Treat Rev* 16: 83–93

5 Combs DW (1995) Recent developments in aromatase inhibitors. *Exp Opin Ther Patents* 5: 529–534

6 Fishman J, Goto J (1981) Mechanisms of oestrogen biosynthesis: participation of multiple enzyme sites in placental aromatase hydroxylations. *J Biol Chem* 256: 4466–4471

7 Means GD, Mahendroo MS, Corbin CJ, Mathis JM, Powell FE, Mendelson CR, Simpson ER (1989) Structural analysis of the gene encoding human aromatase cytochrome P-450, the enzyme responsible for estrogen biosynthesis. *J Biol Chem* 264: 19385–19391

8 Bulun SE, Simpson ER (1994) Regulation of aromatase expression in human tissues. *Breast Cancer Res Treat* 30: 19–29

9 Johnston JO, Metcalf BW (1984) Aromatase: a target enzyme in breast cancer. In: Sunkara P (ed.): *Novel approaches to cancer chemotherapy*. Academic Press, London, 307–328

10 Kao YC, Cam LL, Laughton CA, Zhou D, Chen S (1996) Binding characteristics of seven inhibitors of human aromatase: a site directed mutagenesis study. *Cancer Res* 56: 3451–3461

11 Santen RJ, Misbin RI (1981) Aminoglutethimide: review of pharmacology and clinical use. *Pharmacotherapy* 1: 95–120

12 Vanden Bossche H., Moereels H, Koymans LMH (1994) Aromatic inhibitors -mechanisms for non-steroidal inhibitors. *Breast Cancer Res Treat* 30: 43–55

13 Barone RM, Shamonki IM, Siiteri PK, Judd HL (1979) Inhibition of peripheral aromatization of androstenedionc to estrone in postmenopausal women with breast cancer using A'-testololactone. *J Clin Endocrinol Metab* 49: 672–676

14 Griffiths CT, Hall TC, Saba Z, Barlow JJ, Nevinny HB (1973) Preliminary trial of aminoglutethimide in breast cancer. *Cancer* 32: 31–37

15 Lipton A, Santen RJ (1974) Medical adrenalectomy using aminoglutethimide and dexamethasone in advanced breast cancer. *Cancer* 33: 503–512

16 Segaloff A (1982) Testololactone: clinical trials. *Cancer Res* 42: 3387–3388

17 Segaloff A, Weeth JB, Meyer KK, Rongone EL, Cunningham ME (1962) Hormonal therapy in cancer of the breast. 19. Effect of oral administration of $\Delta^1$-testololactone on clinical course and hormonal excretion. *Cancer* 15: 633–635

18 Newsome HH, Brown PN, Terz JJ, Lawrence W (1977) Medical and surgical adrenalectomy in patients with advanced breast carcinoma. *Cancer* 39: 542–546

19 Hughes SW, Burley DM (1970) Aminoglutethimide. a "side-effect" turned to therapeutic advantage. *Postgrad Med J* 46: 409–416

20 Thompson EA Jr, Siiteri PK (1974) Utilization of oxygen and reduced nicotinamide adenine dinucleotide phosphate by human placental microsomes during aromatization of androstenedione. *J Biol Chem* 249: 5364–5372

21 Santen RJ, Santner S, Davis B, Veldhuis J, Samojlik E, Ruby E (1978) Aminoglutethimide inhibits extraglandular estrogen production in postmenopausal women with breast carcinoma. *J Clin Endocr Metab* 46: 1257–1265

22 Stuart-Harris R, Smith IE, Dowsett M, Bozek T, McKinna JA, Gazet J-C, Jeffcoate SL, Kurkure A, Carr L (1984) Low-dose aminoglutethimide in treatment of advanced breast cancer. *Lancet* 2: 604–607

23 Graves PE, Salhanick HA (1979) Stereoselective inhibition of aromatase by enantiomers of aminoglutethimide. *Endocrinology* 105: 52–57

24 Harris AL, Dowsett M, Cantwell BM, Sainsbury JR, Needham G, Farndon J, Wilson R (1986) Endocrine effects of low dose aminoglutethimide with hydrocortisone – an optimal hormone suppressive regime. *Breast Cancer Res Treat* 7: 68–72

25 Harris AL, Cantwell BM, Sainsbury JR, Needham G, Evans RG, Dawes PT, Wilson R, Farndon J (1986) Low dose aminoglutethimide (125 mg twice daily) with hydrocortisone for the treatment of advanced breast cancer. *Breast Cancer Res Treat* 7: 41–44

26 Murray R, Pitt P (1985) Low-dose aminoglutethimide without steroid replacement in the treatment of postmenopausal women with advanced breast cancer. *Eur J Cancer Clin Oncol* 21: 19–22

27 Harris AL, Powles TJ, Smith IE, Coombes RC, Ford HT, Gazet JC, Harmer GE, Morgan M, White H, Parsons CA et al. (1983) Aminoglutethimide for the treatment of advanced postmenopausal

breast cancer. *Eur J Cancer Clin Oncol* 19: 11–17
28 Gale KE (1982) Treatment of advanced breast cancer with aminoglutethimide: a 14-year experience. *Cancer Res* 42 (suppl.): 3389s–3396s
29 Murray RM, Pitt P (1984) Treatment of advanced metastatic breast cancer, carcinoma of the prostate and endometrial cancer with aminoglutethimide. In: GA Nagel, RJ Santen (eds): *Aminoglutethimide as an aromatase inhibitor in the treatment of cancer*. Hans Huber, Bern, 109–122
30 Lipton A, Santen RJ, Harvey HA (1984) Aminoglutethimide clinical trials. In: HA Harvey, A Lipton, MA Michaels (eds): *Breast cancer: therapeutic modalities current and future*. MES Medical Education Services, Mississauga, 127–135
31 Santen RJ, Samojlik E, Worgul TJ (1981) Aminoglutethimide product profile. In: RJ Santen, IC Henderson (eds): *Pharmanual: a comprehensive guide to the therapeutic use of aminoglutethimide*. S Karger, Basel, 101–160
32 Santen RJ, Worgul TJ, Harvey H, Lipton A, Boucher A, Samojlik E, Wells S (1982) Aminoglutethimide as treatment of postmenopausal women with advanced breast carcinoma. Correlation of clinical and hormonal responses. *Ann Intern Med* 96: 94–101
33 Harris AL (1985) Could aminoglutethimide replace adrenalectomy. *Breast Cancer Res Treat* 6: 201–211
34 Lawrence DV, Lipton A, Harvey HA, Santen RJ, Wells SA, Cox CE, White DS (1980) Influence of estrogen receptor status on response of metastatic breast cancer to aminoglutethimide. *Cancer* 45: 786–791
35 Santen RJ, Worgul TJ, Samojlik E, Interroule A, Boucher AE, Lipton A, Harvey HA, White DS, Smart E, Cox C et al. (1981) A randomised trial comparing surgical adrenalectomy with aminoglutethimide plus hydrocortisone in women with advanced breast cancer. *N Eng J Med* 305: 545–551
36 Harris AL, Dowsett M, Jeffcoate SL, McKenna JA, Morgan M, Smith IE (1982) Endocrine and therapeutic effects of aminoglutethimide in premenopausal patients with breast cancer. *J Clin Endocrinol Metab* 55: 718–720
37 Santen RJ (1984) Panel discussion. In: HA Harvey, A Lipton, MA Michaels (eds): *Breast cancer: therapeutic modalities current and future*. MES Medical Education Services, Mississauga
38 Murray RM, Pitt P (1981) Medical adrenalectomy in patients with advanced breast cancer resistant to anti-oestrogen treatment. *Breast Cancer Res Treat* 1: 91–95
39 Santen RJ, Wells SA (1980) The use of aminoglutethimide in the treatment of patients with metastatic carcinoma of the breast. *Cancer* 46: 1066–1074
40 Smith IE, Harris AL, Morgan M, Ford HT, Gazet JC, Harmer CL, White H, Parsons CA, Villards A, Walsh G et al. (1981) Tamoxifen *versus* aminoglutethimide in the treatment of advanced breast carcinoma. A control randomised cross-over trial. *Br Med J* 283: 1432–1434
41 Smith IE, Harris AL, Morgan M, Gazet JC, McKenna JA (1982) Tamoxifen *versus* aminoglutethimide *versus* combined tamoxifen and aminoglutethimide in the treatment of advanced breast carcinoma. *Cancer Res* 42: 3430–3433
42 Smith IE, Stuart-Harris R, Harris AL, Dowsett M (1984) Aminoglutethimide alone and in combination in the treatment of advanced breast cancer: clinical and endocrine aspects. In: HA Harvey, A Lipton, MA Michaels (eds): *Breast cancer: therapeutic modalities current and future*. MES Medical Education Services, Mississauga, 145–149
43 Stuart-Harris RC, Smith IE (1984) Aminoglutethimide in the treatment of advanced breast cancer. *Cancer Treat Rev* 11: 189–204
44 Brodie AM, Schwarzel WC, Sheikh AA, Brodie HJ (1977) The effect of an aromatase inhibitor, 4-hydroxyandrostenedione, on estrogen-dependent processes in reproduction and breast cancer. *Endocrinology* 110: 1684–1695
45 Schwarzel WC, Kruggel W, Brodie HJ (1973) Studies on the mechanisms of estrogen biosynthesis. 8. The development of inhibitors of the enzyme system in human placenta. *Endocrinology* 92: 866–880
46 Brodie AM, Garnett WM, Hendrickson JR, Tsai-Morris CH, Marcotte PA, Robinson CH (1981) Inactivation of aromatase *in vitro* by 4-hydroxy-4-androstene-3,17-dione and 4-acetoxy-4-androstene-3,17-dione and sustained effects *in vivo*. *Steroids* 38: 693–702
47 Cunningham D, Powles TJ, Dowsett M, Hutchison G, Brodie AM, Ford HT, Gazet JC, Coombes RC (1987) Oral 4-hydroxyandrostenedione, a new endocrine treatment for disseminated breast cancer. *Cancer Chemother Pharmacol* 20: 253–255

48 Dowsett M, Cunningham DC, Stein RC, Evans S, Dehennin L, Hedley A, Coombes RC (1989) Dose-related endocrine effects and pharmacokinetics of oral and intramuscular 4-hydroxyandrostenedione in postmenopausal breast cancer patients. *Cancer Res* 49: 1306–1312
49 MacNeill FA, Jacobs S, Dowsett M, Lonning PE (1995) The effects of oral 4-hydroxyandrostenedione on peripheral aromatisation in post-menopausal breast cancer patients. *Cancer Chemother Pharmacol* 36: 249–254
50 Coombes RC, Goss P, Dowsett M, Gazet JC, Brodie A (1984) 4-Hydroxy-androstenedionein treatment of postmenopausal patients with advanced breast cancer. *Lancet* 2: 1237–1239
51 Dowsett M, Coombes RC (1994) Second generation aromatase inhibitor – 4-hydroxyandrostenedione. *Breast Cancer Res Treat* 30: 81–87
52 Dowsett M, Goss PE, Powles TJ, Hutchinson G, Brodie AMH, Jeffcoate SL, Coombes RC (1987) Use of the aromatase inhibitors 4-hydroxyandrostenedione in postmenopausal breast cancer: optimization of therapeutic dose and route. *Cancer Res* 47: 1957–1961
53 Coombes RC, Hughes SWM, Dowsett M (1992) 4-Hydroxyandrostenedione: a new treatment for postmenopausal patients with breast cancer. *Eur J Cancer* 28: 1941–1945
54 Carrion RP, Candel VA, Calabresi F, Michel RT, Santos R, Delozier T, Goss P, Mauriac L, Feuilhade F, Freue M et al. (1994) Comparison of the selective aromatase inhibitor formestane with tamoxifen as first-line hormonal therapy in postmenopausal women with advanced breast cancer. *Ann Oncol* 5: S19–S24
55 Rose C, Freue M, Kjaer M, Boni C, Janicke F, Coombes C, Willemse PHB, van Belle S, Carrion RP, Jolivet J et al. (1996) An open, comparative randomized trial comparing formestane vs oral megestrol acetate as a second-line therapy in postmenopausal advanced breast cancer patients. *Eur J Cancer* 32A: 49
56 Bhatnagar AS, Hausler A (1987) Fortschritte in der Entwicklung neuer wirksamer und selektiver Aromatasehemmer. In: K Possinger, WR Miller (eds): *Aromatasehemmer Neuer Perspektiven in der Behandlung des Mammakaranoms*. Verlag: Munchen W Zuchshverat, *Aktuelle Onkologie* 38, 23–28
57 Schieweck K, Bhatnager AS, Matter A (1988) CGS 16949A, a new nonsteroidal aromatase inhibitor: effects on hormone-dependent and independent tumors *in vitro*. *Cancer Res* 48: 834–838
58 Santen RJ, Demers LM, Adlercreutz H, Harvey H, Santner S, Sander S, Lipton A (1989) Inhibition of aromatase with CGS16949A in postmenopausal women. *J Clin Endocrinol Metab* 68: 99–106
59 Coombes RC, Stein RC, Dowsett M (1989) Aromatase inhibitors in human breast cancer. In: JS Beck (ed.): *Oestrogen and the human breast.* Proceedings of the Royal Society of Edinburgh, 283–291
60 Santen RJ (1989) Novel methods of estrogen deprivation for treatment of breast disease. In: JS Beck (ed.): *Oestrogen and the human breast*. Proceedings of the Royal Society of Edinburgh.
61 Falkson G, Raats JI, Falkson HC (1992) Fadrozole hydrochloride, a new nontoxic aromatase inhibitor for the treatment of patients with metastatic breast cancer. *J Steroid Biochem Mol Biol* 43: 161–165
62 Hofken K (1993) Experience with aromatase inhibitors in the treatment of advanced breast cancer. *Cancer Treat Rev* 19: 37–44
63 Bonnefoi HR, Smith IE, Dowsett M, Trunet PF, Houston SJ, da Luz RJ, Rubens RD, Coombes RC, Powles TJ (1996) Therapeutic effects of the aromatase inhibitor fadrozole hydrochloride in advanced breast cancer. *Br J Cancer* 173: 539–542
64 Buzdar A, Jonat W, Howell A, Jones SE, Blomqvist C, Vogel CL, Eiermann W, Wolter JM, Azab M, Webster A et al. (1996) Anastrozole, a potent and selective aromatase inhibitor, *versus* megestrol acetate in postmenopausal women with advanced breast cancer: results of overview analysis of two phase III trials. *J Clin Oncol* 14: 2000–2011
65 Thurlimann B, Beretta K, Bacchi M, Castiglione-Gertsch M, Goldhirsch A, Jungi WF, Cavalli F, Senn HJ, Fey M, Lohnert T (1996) First-line fadrozole HCI (CGS 16949A) *versus* tamoxifen in postmenopausal women with advanced breast cancer. Prospective randomised trial of the Swiss Group for Clinical Cancer Research SAKK 20/88. *Ann Oncol* 7: 471–479
66 Lonning PE, Jacobs S, Jones A, Haynes B, Powles TJ, Dowsett M (1991) The influence of CGS 16949A on peripheral aromatisation in breast cancer patients. *Br J Cancer* 63: 789–793
67 Demers LM, Melby JC, Wilson TE, Lipton A, Harvey GA, Santen RJ (1990) The effects of CGS 16949A, an aromatase inhibitor on adrenal mineralocorticoid biosynthesis. *J Clin Endocrinol Metab* 70: 1162–1166

68 Stein RC, Dowsett M, Hedley A, Ford HT, Gazet JC, Coombes RC (1990) Preliminary study of the treatment of advanced breast cancer in postmenopausal women with aromatase inhibitor CGS 16949A. *Cancer Res* 50: 1381–1384
69 Dowsett M, Smithers D, Moore J, Trunet PF, Coombes RC, Powles TJ, Rubens R, Smith IE (1994) Endocrine changes with the aromatase inhibitor fadrozole hydrochloride in breast cancer. *Eur J Cancer* 30A: 1453–1458
70 Plourde PV, Dryoff M, Dukes M (1994) Arimidex (TM): a potent and selective fourth-generation aromatase inhibitor. *Breast Cancer Res Treat* 30: 103–111
71 Demers LM, Lipton A, Harvey HA, Kambic KB, Grossberg H, Brady C, Santen RJ (1993) The efficacy of CGS 20267 in suppressing estrogen biosynthesis in patients with advanced stage breast cancer. *J Steroid Biochem* 44: 687–691
72 Iveson TJ, Smith IE, Ahern J, Smithers DA, Trunet PF, Dowsett M (1993) Phase I study of the oral nonsteroidal aromatase inhibitor CGS 20267 in postmenopausal patients with advanced breast cancer. *Cancer Res* 53: 266–270
73 van der Wall E, Donker TH, de Frankrijker E, Nortier HW, Thijssen JH, Blankenstein RA (1993) Inhibition of the *in vivo* conversion of androstenedione to estrone by the aromatase inhibitor vorozole in healthy postmenopausal women. *Cancer Res* 53: 4563–4566
74 Wouters W, Snoeck E, De Coster R (1994) Vorozole, a specific non-steroidal aromatase inhibitor. *Breast Cancer Res Treat* 30: 89–94
75 Lonning PE (1996) Pharmacology of new aromatase inhibitors. *The Breast* 5: 202–208
76 Geisler J, King N, Dowsett M, Ottestad L, Lundgren S, Walton P, Kormeset PO, Lonning PE (1996) Influence of anastrozole (arimidex), a selective, nonsteroidal aromatase inhibitor, on *in vivo* aromatization and plasma estrogen levels in postmenopausal women with breast cancer. *Br J Cancer* 74: 1286–1291
77 Dowsett M (1996) Biological background to aromatase inhibition. *The Breast* 5: 196–201
78 Demers LM (1994) Effects of Fadrozole (CGS 16949A) and Letrozole (CGS 20267) on the inhibition of aromatase activity in breast cancer patients. *Breast Cancer Res Treat* 30: 95–102
79 Bhatnagar AS, Hausler A, Schieweck K, Lang M, Bowman R (1995) Highly selective inhibition of estrogen biosynthesis by CGS 20267, a new non-steroidal aromatase inhibitor. *J Steroid Biochem Mol Biol* 37: 1021–1027
80 Geisler J, Haynes B, Anker G, Dowsett M, Lonning PE (2002) Influence of letrozole and anastrozole on total body aromatization and plasma estrogen levels in postmenopausal breast cancer patients evaluated in a randomized, cross-over study. *J Clin Oncol* 20: 751–757
81 Gershanovich M, Chaudri HA, Campos D, Lurie H, Bonaventura A, Jeffrey M, Buzzi F, Bodrogi I, Ludwig H, Reichardt P et al. (1998) Letrozole, a new oral aromatase inhibitor: randomized trial comparing 2.5 mg daily, 0.5 mg daily and aminoglutethimide in postmenopausal women with advanced breast cancer. *Ann Oncol* 9: 639–645
82 Johannessen DC, Engan T, Di Salle E, Zurlo MG, Paolini J, Ornati G, Piscitelli G, Kvinnsland S, Lonning PE (1997) Endocrine and clinical effects of exemestane (PNU 155971), a novel steroidal aromatase inhibitor, in postmenopausal breast cancer patients: a phase I study. *Clin Cancer Res* 3: 1101–1108
83 Vermeulen A, Paridaens R, Heuson JC (1983) Effects of aminoglutethimide on adrenal steroid secretion. *Clin Endocrinol* 19: 673–682
84 Miller WR, O'Neill JS (1989) The relevance of local estrogen metabolism within the breast. *Proc Roy Soc Edin* 95B: 203–217
85 Simpson ER, Mendelson CR (1989) The regulation of estrogen biosynthesis in human adipose tissue. In: *Oestrogen and the human breast. Proc Roy Soc Edin* 95: 153–159
86 Samojlik E, Santen RJ, Worgul TJ (1984) Suppression of residual oestrogen production with aminoglutethimide in women following surgical hypophyscctomy or adrenalectomy. *Clin Endocrinol* 20: 43–51
87 Adams JB, Garcia M, Rochefort H (1981) Estrogenic effects of physiological concentrations of 5-androstene 3B17B diol and its metabolism in MCF-7 human breast cancer cells. *Cancer Res* 41: 4720–4726
88 Wander HK, Blosscy HCh and Nagcl GA (1986) Aminoglutethimide in the treatment of premenopausal patients with metastatic breast cancer. *Eur J Cancer Clin Oncol* 22: 1371–1374
89 Santen RJ, Samojlik E, Wells SA (1980) Resistance of the ovary to blockade of aromatization with aminoglutethimide. *J Clin Endocrinol Metab* 51: 473–477
90 Stein RC, Dowsett M, Hedley A, Gazet JC, Ford HT, Coombes RC (1990) The clinical end

endocrine effects of 4-hydroxyandrostenedione alone and in combination with goserelin in premenopausal women with advanced breast cancer. *Br J Cancer* 52: 679–683
91 Santen RJ, Manni A, Harvey H, Redmond C (1990) Endocrine treatment of breast cancer in women. *Endocr Rev* 11: 1–45
92 Sherman BM, Chapler FK, Crickard K, Wycoff D (1979) Endocrine consequences of continuous antiestrogen therapy with tamoxifen in premenopausal women. *J Clin Invest* 64: 398–404
93 Groom GV, Griffiths K (1976) Effect of the anti-oestrogen tamoxifen on plasma levels of luteinizing hormone, follicle stimulating hormone, prolactin, oestradiol and progesterone in normal premenopausal women. *J Endocrinol* 70: 421–428
94 Liehr JG (2002) Breast carcinogensis and its prevention by inhibition of estrogen genotoxicity. In: WR Miller, JN Ingle (eds): *Endocrine therapy in breast cancer*. Marcel Dekker, New York, 287–301
95 Sharpe RM (1994) Could environmental oestrogenic chemicals be responsible for some disorders of human male reproductive development. *Curr Opin Urol* 4: 295–301
96 Setchell KD, Borriello SP, Hulme P, Kirk DN, Axelson M (1984) Nonsteroidal estrogens of dietary origin: possible roles in hormone-dependent disease. *Am J Clin Nutrition* 40: 569–578
97 McLachlan JA, Newbold R (1987) Estrogens and development. *Environ Health Perspect* 75: 25–27
98 Mouridsen H, Gershanovich M, Sun Y, Perez-Carrion R, Boni C, Monnier A, Apffelstaedt J, Smith R, Sleeboom HP, Janicke F et al. (2001) Superior efficacy of letrozole *versus* tamoxifen as first-line therapy for postmenopausal women with advanced breast cancer: results of a Phase III study of the International Letrozole Breast Cancer Group. *J Clin Oncol* 19: 2596–2606
99 Mouridsen H, Sun Y, Gershanovich M, Perez-Carrion R, Smith R, Chaudri-Ross HA, Lang R, Brady C, Dugan M (2001) Final survival analysis of the double-blind, randomized, multinational Phase III trial of letrozole (Femara®) compared to tamoxifen as first-line hormonal therapy for advanced breast cancer. *Breast Cancer Res Treat* 69: 211
100 Nabholtz JM, Buzdar A, Pollak M, Harwin W, Burton G, Mangalik A, Steinberg M, Webster A, von Euler M (2000) Anastrozole is superior to tamoxifen as first-line therapy for advanced breast cancer in postmenopausal women: results of a North American multicenter randomized trial. Arimidex Study Group. *J Clin Oncol* 18: 3758–3767
101 Dixon JM, Love CD, Bellamy CO, Cameron DA, Leonard RC, Smith H, Miller WR (2001) Letrozole as primary medical therapy for locally advanced and large operable breast cancer. *Breast Cancer Res Treat* 66: 191–199
102 Eiermann W, Paepke S, Appfelstaedt J (2001) Letrozole Neoadjuvant Breast Cancer Study Group. Preoperative treatment of postmenopausal breast cancer patients with letrozole. A randomized double-blind multicenter study. *Ann Oncol* 12: 1527–1532
103 ATAC Trialist's Group (2002) Anastrozole alone or in combination with tamoxifen *versus* tamoxifen alone for adjuvant treatment of postmenopausal women with early breast cancer: first results of the ATAC randomised trial. *Lancet* 359: 2131–2139
104 Miller WR, Sharpe RM (1998) Environmental oestrogens and human reproductive cancers. *Endocr Relat Cancer* 5: 69–96
105 Miller WR, Dixon JM, Macfarlane L, Cameron D, Anderson TJ (2002) Pathological features of breast cancer response following neoadjuvant treatment with either letrozole or tamoxifen. *Eur J Cancer* 39: 462–468
106 Lipton A, Demers LM, Harvey HA, Kambic KB, Grossberg H, Brady C, Adlercruetz H, Trunet PF, Santen RJ (1995) Letrozole (CGS 20267). A Phase I study of a new potent oral aromatase inhibitor of breast cancer. *Cancer* 75: 2132–2138
107 Johnston SR, Haynes BP, Sacks NP, McKinna JA, Griggs LJ, Jarman M, Baum M, Smith IE, Dowsett M (1993) Effect of oestrogen receptor status and time on the intra-tumoural accumulation of tamoxifen and N-desmethyltamoxifen following short-term therapy in human primary breast cancer. *Breast Cancer Res Treat* 28: 241–250
108 Miller WR, Anderson TJ, Iqbal S, Dixon JM (2002) Neoadjuvant therapy: prediction of response. In: WR Miller, JN Ingle (eds): *Endocrine therapy in breast cancer*. Marcel Dekker, New York, 223–229
109 Ellis MJ, Coop A, Singh B, Mauriac L, Llombert-Cussac A, Janicke F, Miller WR, Evans DB, Dugan M, Brady C (2001) Letrozole is more effective neoadjuvant endocrine therapy than tamoxifen for ErbB-1- and/or ErbB-2-positive, estrogen receptor-positive primary breast cancer: evidence from a phase III randomized trial. *J Clin Oncol* 19: 3808–3816

110 Dixon JM, Jackson J, Hills M, Renshaw L, Cameron DA, Anderson TJ, Miller WR, Dowsett M (2004) Anastrozole demonstrates clinical and biological effectiveness in oestrogen receptor-positive breast cancers, irrespective of the erbB2 status. *Eur J Cancer* 40: 2742–2747
111 Allegra JC, Barlock A, Huff KK, Lippman ME (1980) Changes in multiple or sequential estrogen receptors in breast cancer. *Cancer* 45: 792–794
112 Hawkins RA, Tesdale AL, Anderson ED, Levack PA, Chetty U, Forrest AP (1990) Does the oestrogen receptor concentration of a breast cancer change during systemic therapy? *Br J Cancer* 61: 877–880
113 Miller WR, Hawkins RA, Mullen P, Sourdaine P, Telford J (1995) Aromatase inhibition: determinants of response and resistance. *Endocr Relat Cancer* 2: 73–85
114 Fuqua SA, Wiltschke C, Castles C, Wolf D, Allred DC (1995) A role for estrogen-receptor variants in endocrine resistance. *Endocr Relat Cancer* 2: 19–25
115 Fujimoto N, Katzenellenbogen BS (1994) Alteration in the agonist/antagonist balance of antiestrogens by activation of protein kinase A signalling pathways in breast cancer cells: antiestrogen-selectivity and promoter-dependence. *Mol Endocrinol* 8: 296–304
116 Isaacs JT (1988) Clonal heterogeneity in relation to response. In: BA Stoll (ed.): *Endocrine management of cancer: biological bases*. Karger, Basel, 125–140
117 Howell A, Defriend D, Anderson E (1995) Clues to the mechanism of endocrine resistance from clinical studies in advanced breast cancer. *Endocr Relat Cancer* 2: 131–139
118 Santen RJ (1982) Overall experience with aminoglutethimide in the management of advanced breast cancer. In: RW Elsdon-Dew, IM Jackson, GFB Birdwood (eds): *Aminoglutethimide: an alternative endocrine therapy for breast carcinoma*. Academic Press, London, 3–7
119 Herman ME, Katzenellenbogen B (1994) Alterations in transforming growth factor-$\alpha$ and -$\beta$ production and cell responsiveness during the progression of MCF-7 human breast cancer cells to estrogen-autonomous growth. *Cancer Res* 54: 5867–5874
120 King RJ, Wang DY, Daly RJ, Darbre PD (1989) Approaches to studying the role of growth factors in the progression of breast tumours from the steroid sensitive to insensitive state. *J Steroid Biochem* 34: 133–138
121 Miller WR, O'Neill JS (1988) The importance of local synthesis of estrogen within the breast. *Steroids* 50: 537–548
122 Kadohama N, Yarborough C, Zhou D, Chen S, Osawa Y (1992) Kinetic properties of aromatase mutants ProSOSPhe, Asp309Asn and Asp309Ala and their interactions with aromatase inhibitors. *J Steroid Biochem Mol Biol* 43: 693–701
123 James VH, Reed MJ, Adams EF, Ghilchick M, Lai LC, Coldham NG, Newton CJ, Purohit A, Owen AM, Singh A et al. (1989) Oestrogen uptake and metabolism *in vivo*. *Proc Roy Soc Edin* 95B: 185–193
124 Miller WR (1992) *In vitro* and *in vivo* effects of 4-hydroxyandrostenedione on steroid and tumour metabolism. In: RC Coombes, M Dowsett (eds): *4-Hydroxy-androstenedione – a new approach to hormone-dependent cancer*, International Congress and Symposium Series. Royal Society of Medicine Services, London, 45–50
125 Sourdaine P, Parker MG, Telford J, Miller WR (1994) Analysis of the aromatase cytochrome P450 gene in human breast cancer. *J Mol Endocrinol* 13: 331–337
126 Lonning PE, Bajetta E, Murray R, Tubiana-Hulin M, Eisenberg PD, Mickiewicz E, Celio L, Pitt P, Mita M, Aaronson NK et al. (2000) Activity of exemestane (Aromasin) in metastatic breast cancer after failure of nonsteroid aromatase inhibitors: a phase II trial. *J Clin Oncol* 18: 2234–2244
127 Carlini P, Frassoldati A, De Marco S, Casali A, Ruggeri EM, Nardi M, Papaldo P, Fabi A, Paoloni F, Cognetti F (2001) Formestane, a steroidal aromatase inhibitor after failure of non-steroidal aromatase inhibitors (anastrozole and letrozole): is a clinical benefit still available? *Ann Oncol* 12: 1539–1543
128 Miller WR, Jackson J (2003) The therapeutic potential of aromatase inhibitors. *Expert Opin Invest Drugs* 12: 337–351
129 Goss PE (2001) Chemoprevention with aromatase inhibitors. In: WR Miller, RJ Santen (eds): *Aromatase inhibition and breast cancer*. Marcel Dekker, New York, 161–181
130 Kaplowitz PB (2001) Aromatase inhibitors as therapy for pubertal gynecomastia. In: WR Miller, RJ Santen (eds): *Aromatase inhibition and breast cancer*. Marcel Dekker, New York, 259–266
131 Smith MR (2001) Aromatase inhibition and prostate cancer. In: WR Miller, RJ Santen (eds): *Aromatase inhibition and breast cancer*. Marcel Dekker, New York, 271–276

132 Bulun S, Zeitoun KM, Takayama K, Sasano H, Simpson ER (2001) Aromatase in endometriosis: biological and clinical application. In: WR Miller, RJ Santen (eds): *Aromatase inhibition and breast cancer*. Marcel Dekker, New York, 279–291

133 Geisler J, King N, Anker G, Ornati G, Di Salle E, Lønning PE, Dowsett M (1998) *In vivo* inhibition of aromatization by exemestane, a novel irreversible aromatase inhibitor, in postmenopausal breast cancer patients. *Clin Cancer Res* 4 (9): 2089–2093

Aromatase Inhibitors
Edited by B.J.A. Furr



# Aromatase inhibitors and models for breast cancer

Angela Brodie

*Department of Pharmacology & Experimental Therapeutics, University of Maryland, School of Medicine, Baltimore, MD 21201, USA*

## Introduction

Two approaches that are used to ameliorate the growth effects of oestrogens on primary and metastastic breast cancers are the inhibition of oestrogen action by compounds interacting with oestrogen receptors (ERs; antioestrogens) and the inhibition of oestrogen synthesis by inhibitors of the enzyme, aromatase. Treatment with the antioestrogen, tamoxifen, has been an important therapeutic advance in breast cancer management for patients with ER-positive tumours. However, concerns exist about the long-term use of this antioestrogen. Although tamoxifen functions as an ER antagonist, it also exhibits weak or partial agonist properties. The antioestrogenic activity of tamoxifen is limited to its effects on breast tumour cells whereas in other regions of the body tamoxifen may actually function as an oestrogen agonist. This can lead to increased risk of hyperplasia of the endometrium and occasionally cancer and increased risk of strokes [1, 2]. These agonist effects of tamoxifen were realized from its inception [3]. Because of these concerns, we proposed selective inhibition of aromatase to reduce oestrogen production as a different strategy that is unlikely to be associated with oestrogenic effects. For this reason, aromatase inhibition could have greater antitumour efficacy than tamoxifen. The selective approach would not interfere with other cytochrome P450 enzymes involved in the synthesis of essential hormones such as cortisol and aldosterone. Thus, selective aromatase inhibition would be a safer and more effective approach than antioestrogens. A number of compounds that are selective inhibitors of aromatase were first reported in 1973 [4].

## Model systems for studying aromatase inhibitors *in vitro*

During pregnancy, the placenta expresses high levels of aromatase in the syncytiotrophoblasts in the outer layer of the chorionic villi [5, 6] and is an excellent source of highly active enzyme [4, 7]. Placental microsomes have been used to study aromatase since the 1950s. The conversion of radiolabeled substrate androstenedione to oestrogen in the presence of candidate inhibitors after incubation with human placental microsomes proved a valuable system for identifying compounds as aromatase inhibitors.

Following the initial publication of Brodie and colleagues [4, 8, 9], a number of groups reported novel steroidal compounds as inhibitors of aromatase during the late 1970s and 1980s. These steroid analogues showed competitive inhibition kinetics. However, further studies revealed that several steroidal inhibitors, notably 4-hydroxyandrostenedione (4-OHA), 4-acetoxy-A [10, 11], 1,4,6-androstatriene-3,17-dione (ATD), A-trione, 10β-propargyloest-4-ene-3,17-dione (10-PED) [12–14], 16-brominated androgen derivatives [15], and 7α-*p*-amino-thiophenyl-androstenedione [16–18], also cause time-dependent loss of aromatase activity in placental microsomes when pre-incubated in the absence of substrate, but in the presence of NADPH. No loss of enzyme activity occurred without added cofactors. These findings suggest that steroidal inhibitors can cause long-term inactivation (or irreversible inhibition) of aromatase. Studies with exemestane demonstrate that this steroidal inhibitor also causes aromatase inactivation [19, 20].

Siiteri and Thompson [21, 22] tested a series of known compounds as aromatase inhibitors in placental microsomes. Of these, testololactone, a steroidal compound that has been used for some 20 years in breast cancer therapy, and aminoglutethimide were reported by them to inhibit aromatization. Testololactone had rather weak activity, but aminoglutethimide was an effective aromatase inhibitor. Originally used to inhibit adrenal steroidogenesis in breast cancer patients [23], its use as an aromatase inhibitor contributed to establishing a place for aromatase inhibition in breast cancer treatment [24]. This compound interferes with cytochrome P450 and therefore inhibits aromatase as well as 20α-, 18-, and 11β-hydroxylases [25].

Following several years of preclinical development [8, 26, 27], the first selective inhibitor, formestane (4-OHA; lentaron), was evaluated clinically and was found to be effective for the treatment of breast cancer [28, 29]. As indicated above, formestane is a substrate analogue and mechanism-based inhibitor (suicide inhibitor) that inactivates the enzyme by binding irreversibly [10, 11]. Subsequently, exemestane (aromasin) became available and is also in this class of inhibitors.

A number of non-steroidal aromatase inhibitors were later developed and include the highly potent triazole compounds letrozole and anastrozole. Non-steroidal inhibitors possess a heteroatom such as a nitrogen-containing heterocyclic moiety. This interferes with steroidal hydroxylation by binding with the haem iron of cytochrome P450 arom. These compounds are reversible inhibitors of aromatase. Most non-steroidal inhibitors are intrinsically less enzyme-specific and will inhibit, to varying degrees, other cytochrome P450-mediated hydroxylations in steroidogenesis. However, anastrozole and letrozole are highly selective for aromatase. Good specificity and potency are important determinants in achieving drugs with few side effects. Both classes of inhibitors, steroidal enzyme inactivators and non-steroidal triazole compounds, have proved to be well-tolerated agents in clinical studies. The two triazole inhibitors, letrozole and anastrozole, as well as exemestane, are now approved in the USA for breast cancer treatment [30]. Recent studies have shown that these aromatase inhibitors are more effective than tamoxifen [31–35].

## Model systems for studying aromatase and aromatase inhibitors *in vivo*

### *Determining inhibition of oestrogen synthesis and production*

When active inhibitors had been identified in human placental microsomes, studies in animal models were essential to define the ability of the compounds to inhibit oestrogen production *in vivo*. For this purpose, a number of rodent and non-human primate models were developed. These include models to determine the effects of an inhibitor on oestrogen production and the endocrine system, as well as the antitumour efficacy of the compound.

*Pregnant mare's serum gonadotrophin (PMSG)-primed rat model*
To determine whether aromatase inhibitors would inhibit oestrogen synthesis and production *in vivo*, rats primed for 12 days previously with PMSG to stimulate aromatase activity and maintain a constant oestrogen output were employed in early studies of formestane (4-OHA) and other inhibitors [36, 37]. The value of this model was to demonstrate that aromatase inhibitors reduce oestrogen secretion *in vivo* by direct inhibition of ovarian aromatization rather than by other mechanisms that might cause reduction in oestrogen levels. In this model, it is unlikely that oestrogen production would be suppressed by compounds acting mainly by negative feedback on luteinizing hormone (LH) and follicle-stimulating hormone (FSH) secretion, since PMSG injections would override potential changes in endogenous gonadotrophins.

In this model, oestrone production is measured in ovarian vein blood collected by cannulation and aromatase activity is measured in ovarian microsomes prepared at various times after the injection. In studies of 4-OHA, 24 h after injection, ovarian aromatase activity was reduced and remained suppressed even up to 72 h. Oestrogen concentrations measured by radioimmunoassay in the ovarian vein blood were also much reduced by inhibitor treatment. Additional information gained from studies with the PMSG-primed rat is the specificity of the candidate compound for oestrogen biosynthesis. Thus no significant difference was found between the concentrations of progesterone, testosterone, or androstenedione in peripheral plasma of control rats and plasma collected 3 h after injection of 4-OHA, indicating that the main action of this compound was on aromatase.

*Normal cycling rats*
When aromatase inhibitors were administered to female rats early in the oestrous cycle, the sequence of events leading to ovulation was inhibited. In addition, when rats were injected on the morning of pro-oestrus (11:00 h) with

inhibitor (50 mg/kg) ovulation could also be inhibited. Thus, 3 h after injection, at the time that the normal oestrogen peak occurs, blood was collected by ovarian vein cannulation for oestrogen determinations. Oestrogen secretion was reduced, the preovulatory LH surge was inhibited, and ovulation prevented [37]. When oestradiol was given in addition to aromatase inhibitor treatment, these effects were reversed and mating occurred at the normally expected times, indicating that the lack of ovulation during inhibitor treatment was the result of reduced oestrogen secretion. This model also provided information on the effect of inhibiting oestrogen on ovulation.

### *Aromatase-knockout model*

Knowledge concerning the effects of oestrogens on different target tissues has been provided using disruption of the aromatase and ER gene (knockout models). Several models have been developed that include the aromatase-knockout mouse (ArKO) [38], the ERKO mouse (disrupted ER-$\alpha$), the $\beta$ERKO mouse (disrupted ER-$\beta$), as well as the $\alpha/\beta$ERKO-mouse (disrupted ER-$\alpha$ and ER-$\beta$) [39]. These model systems are valuable for studying the function of aromatase and the individual ERs *in vivo*.

### *Int-5/aromatase model*

A model that has been valuable for investigating the role of oestrogen in breast cancer is the int-5/aromatase transgenic mouse developed by Tekmal and colleagues [40]. Aromatase overexpression contributes to increased oestrogenic activity in the mouse mammary gland, resulting in hyperplastic, dysplastic, and several premalignant changes. These changes persist for several months after post-lactational involution and occur even without circulating ovarian oestrogens in ovariectomized mice, indicating that more than one event is required for tumour formation. These changes can be abrogated by aromatase inhibitors. Thus, early oestrogen exposure of mammary epithelial cells leads to preneoplastic changes, increases susceptibility to environmental carcinogens, and may result in acceleration and/or an increase in the incidence of breast cancer. In male aromatase-transgenic mice [41, 42] the induction of gynecomastia and testicular cancer suggests that tissue oestrogens play a direct role in mammary tumourigenesis. Consistent with these findings, studies by Fisher et al. [38], have shown that oestrogen deficiency in aromatase-knockout mice leads to underdeveloped genitalia and immature mammary glands. Although the mammary glands of female aromatase-transgenic mice exhibited hyperplastic and dysplastic changes, palpable mammary tumours have not been observed even in animals more than 2 years old. This suggests that other cooperating factor(s) or carcinogenic events are required for development of cancer. Thus administration of a single dose of dimethyl-benzanthracene

(DMBA) resulted in the induction of frank mammary tumours in about 25% of aromatase-transgenic mice, and all animals had microscopic evidence of tumour formation, whereas there was no evidence of tumours in DMBA-treated non-transgenic mice [43]. These observations suggest that locally produced oestrogen increases susceptibility to environmental carcinogens.

## *Models for determining antitumour efficacy*

### *Rat model with carcinogen-induced hormone-dependent mammary tumours*

Mammary tumours induced in the female Sprague–Dawley rat with the carcinogen DMBA or nitrosomethyl urea (NMU) have been widely used for studying hormone-dependent tumour growth and the effects of aromatase inhibitors [8, 27, 44, 45] as well as antioestrogens [46, 47]. In this model, tumour growth is dependent on oestrogen produced by the rat ovaries where aromatase is under the control of FSH. Regulation of aromatase gene expression is tissue-specific via 10 promoters spliced into exons; promoter II.2 is the one primarily regulating aromatase in the ovary.

Although rats rarely develop mammary tumours, animals administered DMBA (20 mg/2 ml) by gavage when they are between 50 and 55 days of age will develop tumours in approximately 6–8 weeks [48]. Multiple superficial mammary tumours are induced but do not metastasize. About 80–90% of these tumours are hormone-dependent. Tumours are measured with calipers and their volumes calculated [49]. Groups of rats, for treatment *versus* control studies, are matched as closely as possible for numbers of animals and tumours and for total tumour volumes at the start of the experiment.

Early experiments with 4-OHA [8], 4-acetoxy-A, and ATD [44, 45] in the DMBA model (Fig. 1) showed marked regression of mammary tumours after 4 weeks of treatment. Over 90% of tumours regressed to less than half their original size with 4-acetoxy-A, ATD, and 4-OHA. By contrast, two other aromatase inhibitors, testololactone (Teslac) [50] and aminoglutethimide [51], were much less effective in these experiments [27]. There was no significant tumour regression with testololactone (25 mg/kg per day) compared with controls. With aminoglutethimide injections (25 mg/kg per day), tumour growth was less than controls, but there was no decrease in the percentage change in the total tumour volume.

In this rat model system, 4-OHA and 4-acetoxy-A in comparison to and in combination with tamoxifen (ICI 46,474) were found to be more effective in causing mammary tumour regression when used alone [27]. At the end of 4-week aromatase inhibitor treatment, blood was collected for steroid radioimmunoassay from the ovarian veins of rats with DMBA-induced tumours. Tamoxifen was found to increase oestrogen secretion and to be partially oestrogenic. Other workers have observed similar effects of tamoxifen [52]. The latter property may be responsible for retarding the full effect of the aromatase inhibitor when used in combination with tamoxifen [27].

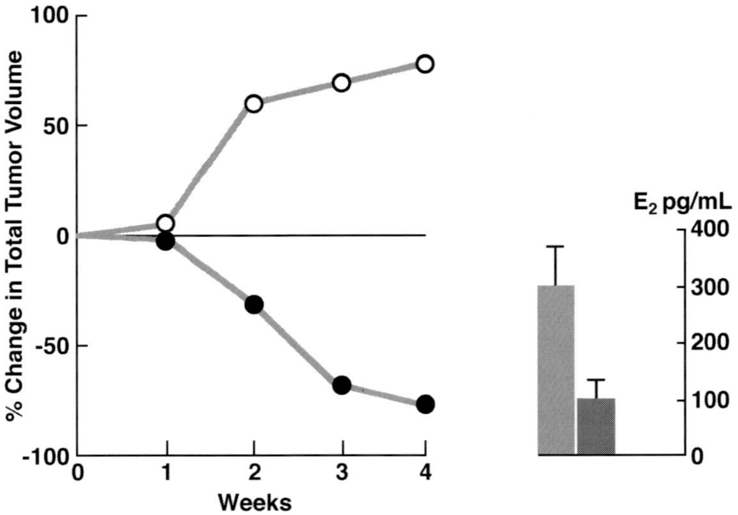


Figure 1. The effect of 4-OHA on DMBA-induced, hormone-dependent mammary tumours of the rat. ●, Percentage change in total volume of 13 tumours on six rats injected with 4-OHA (50 mg/kg per day), twice daily for four weeks; ○, tumours on five control rats injected twice daily with vehicle. At the end of treatment blood was collected from each rat by ovarian vein cannulation for oestradiol ($E_2$) assay; controls were sampled during dioestrus.

*Aromatase inhibitor effects on gonadotrophins*

Secretion of both oestrone and oestradiol was reduced by aromatase inhibitor to below basal values of control rats sampled on oestrus or dioestrus. Trunk blood was collected at autopsy from the aromatase inhibitor-treated rats with DMBA-induced tumours for assay of LH, FSH, and prolactin. Although oestrogen secretion was reduced with 4-acetoxy-A, gonadotrophin concentrations were found to be similar to basal control values, suggesting there may be a direct effect on gonadotrophins. Furthermore, when ovariectomized rats were treated with inhibitors, the rise in LH and FSH that usually occurs in castrates was prevented [27]. Subsequent studies suggested that 4-OHA seems to affect gonadotrophins and aromatase with about equal potency *in vivo*. Since FSH is known to be involved in regulating ovarian aromatase, maintaining basal gonadotrophin concentrations would contribute to the effectiveness of 4-OHA in reducing oestrogen production.

4-OHA and aminoglutethimide decreased ovarian aromatase activity and oestrogen secretion to a similar extent in acute experiments in which rats were given injections on the morning of pro-oestrus, and tissues and blood were collected 3 h later [27]. However, in long-term experiments of 2 and 4 weeks, it is evident that oestradiol suppression was not maintained by aminoglutethimide to the same degree. The initial 90% inhibition of ovarian oestradiol synthesis by aminoglutethimide leads to increased LH levels through feedback-regulatory mechanisms in the intact rat. Reflex increases in LH and FSH were observed in premenopausal patients treated with aminoglutethimide [53]. Thus increased gonadotrophins may tend to stimulate aromatase synthesis by the ovaries and counteract the inhibitory effects of aminoglutethimide to some extent. After 2 weeks in the normal cycling animals, there was a 50% reduction in the mean value of ostradiol that, due to variation, was not significantly different from the control value. Moreover, after 4 weeks of treatment, oestradiol production in five out of six tumour-bearing animals was within the range of values for control animals. This amount of oestradiol was sufficient to maintain the uterine weight comparable to intact control rats. Aminoglutethimide appeared to have no direct effect on either the uterus or pituitary gland in ovariectomized rats, whereas marked reduction in LH levels by 4-OHA suggests a direct action of this compound independent of aromatase inhibition. The effect on LH secretion as well as on the uterus appears to be due to weak androgenic activity (<1% testosterone) of 4-OHA [54] that may contribute to its efficacy in causing regression of DMBA-induced mammary tumours. Thus 4-OHA by more potent aromatase inhibition and gonadotrophin suppression may prevent new enzyme synthesis and follicular development by the ovary, resulting in a greater and sustained reduction in oestradiol production than aminoglutethimide. Whereas these models provided important information about the effects of aromatase inhibitors, it became apparent that maintaining inhibition of ovarian oestrogen production is required for successful treatment in premenopausal patients with hormone-dependent breast cancer. To date, most clinical studies have focused on investigating aromatase inhibitors in postmenopausal patients.

## Models for postmenopausal breast cancer

A large proportion of breast cancer patients are postmenopausal women with ER-positive tumours responsive to hormone therapy. Following the menopause, adipose tissue is considered to be the main site of oestrogen synthesis contributing to circulating oestrogen levels [55]. However, breast tissue has been found to have several-fold higher levels of oestrogen than those in plasma of postmenopausal patients [56–58]. A number of reports indicate that aromatase mRNA as well as aromatase activity is present in normal breast tissue and breast tumours [59–65]. Approximately 60% of breast tumours express aromatase [63] and have aromatase activity [66]. Aromatase expression in extra-gonadal sites is not regulated by FSH but by glucocorticoids, cAMP, prostaglandin $PGE_2$, and other factors. In breast cancer, prostaglandin $PGE_2$, the product of the inducible form of cyclooxygenase (COX-2), appears to be an important mediator of aromatase expression [67, 68]. Thus in postmenopausal breast cancer patients, oestrogen synthesis is independent of feedback regulation between the pituitary gland and the ovary. As mentioned above, the tissue-specific manner of aromatase regulation involves the use of alternative promoters [69]. In peripheral tissue, two promoters, promoters II and 1.3, regulate the enzyme [69, 70].

*JEG-3 tumours demonstrate aromatase inhibition*

A model utilized to demonstrate inhibition of non-ovarian aromatase *in vivo* was introduced by Johnston et al. [71], who employed the athymic, immune-suppressed mouse with tumours grown from human choriocarcinoma cells. Both JEG and JAR cell lines express high levels of aromatase. However, the tumours are not dependent on oestrogens to stimulate their growth. In this model, the aromatase inhibitor 10-PED demonstrated almost complete inhibition of oestrogen production [71].

*Model for peripheral aromatization*

Measurement of *in vivo* peripheral aromatization is an important indicator in determining efficacy of aromatase inhibitors. In early preclinical studies of aromatase inhibitors, the male rhesus monkey was used as a model for determining peripheral aromatization (Fig. 2) [26]. This species was selected

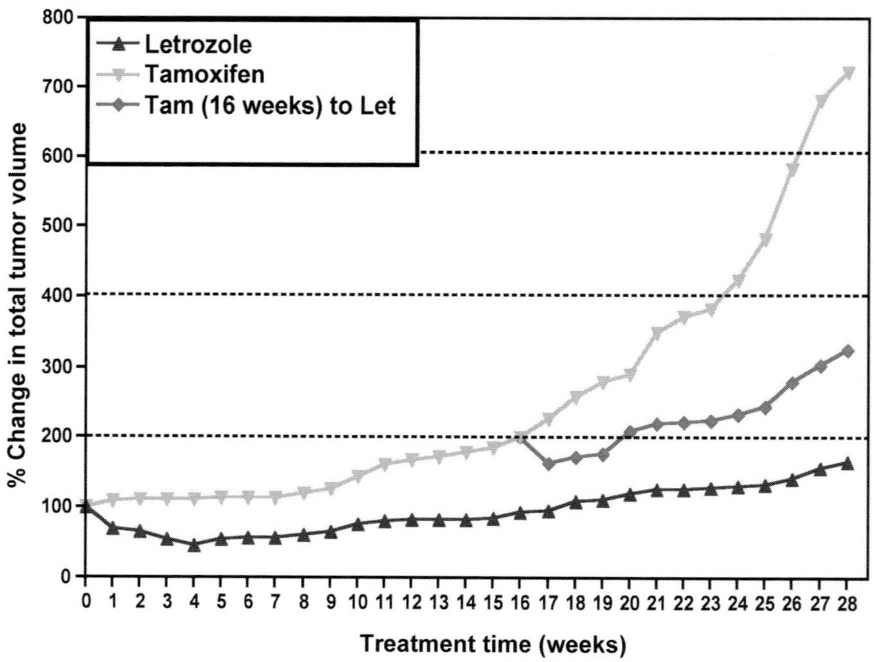


Figure 2.The effect of second-line treatment with letrozole (Let) on the growth of MCF-7Ca breast cancer xenograft tumours progressing on tamoxifen (Tam) treatment. Tumours in the mice treated with tamoxifen (100 μg/day) doubled in volume after 16 weeks of treatment. At that point, the mice were divided into three groups: for continued treatment with tamoxifen ($n = 4$), for second-line treatment with letrozole (10 μg/day; $n = 5$), and for continued treatment with letrozole ($n = 5$). Second-line treatment lasted for 12 weeks, and tumour volumes were measured weekly for a total of 28 weeks. Tumour volumes are expressed as the percentage change relative to the initial tumour volume. Letrozole was not as effective as a second-line treatment as it was as a first-line treatment [80].

because it had been found previously to be a useful model for studying androgen and oestrogen metabolism and dynamics [72]. Similar methodology was used by Santen and colleagues [24] in breast cancer patients to study inhibition of oestrogen production by aminoglutethimide. Recent studies by Lonning et al. [73] suggest that the potency of inhibitors of peripheral aromatase correlates with clinical outcome in patients.

To measure peripheral aromatization, each monkey was infused with [7-$^{3}$H]androstenedione and [4-$^{14}$C]oestrone at a constant rate via the brachial vein. Blood samples were drawn from the femoral vein during infusion at 0, 2.5, 3, and 3.5 h, and steady-state conditions were verified. The conversion of androstenedione to estrone was measured in the samples. Four of the monkeys were treated with injections of 4-OHA (50 mg/kg) at 5 pm on the day before infusion of radiolabeled androstenedione and 1.5 h before beginning the infusion. Each animal served as its own control, being injected with vehicle at the above times before infusion: two monkeys had control infusions 1 week before and two monkeys 1 week after 4-OHA treatment.

Silastic wafers containing 4-acetoxy-A were implanted into two other monkeys 24 h before infusion. Each was also injected with 4-acetoxy-A at 9 am and 5 pm on the day before infusion and 15–30 min before infusion began. Control infusions were performed 1 month after 4-acetoxy-A treatment. Interestingly, peripheral aromatization was very low in the control infusions performed 1 month after treatment, suggesting sustained effects of treatment possibly due to inactivation of aromatase by this steroidal inhibitor. Aromatization rates were reduced by up to 97% of control values. Additional analysis of the samples revealed no specific effects on the metabolic clearance rates of androstenedione and oestrone, the interconversion of the androgens or oestrogens, or on androstenedione conversion to dihydrotestosterone.

### *The mouse xenograft model*

In order to study the antitumoural effects of hormonal agents such as aromatase inhibitors and antioestrogens, a xenograft model was developed that simulates the physiology of the menopausal patient [74, 75]. The athymic, immune-suppressed mouse [76] with tumours grown from human ER-positive breast carcinoma cells (MCF-7) has been used extensively for studies of antioestrogens [46, 47, 77]. As these cells grow rather poorly in intact mice, ovariectomized animals supplemented with oestradiol are usually used. This has the advantage of resembling the physiology of the postmenopausal woman in that oestrogen is available to stimulate tumour proliferation from a non-ovarian source not under gonadotrophin feedback regulation. While the athymic mouse with MCF-7 tumours proved to be an excellent model for studying antioestrogens, it is not useful for investigating the effects of reducing oestrogen production with aromatase inhibitors since MCF-7 cells express only low levels of aromatase [78]. As indicated above, a number of studies have demonstrated that in humans oestrogens are produced locally by aromatase within the breast and by the tumour [65]. In order to replicate this situation in the mouse model, we have utilized MCF-7 cells stably transfected with aromatase (MCF-$7_{CA}$) [74]. As the rodent has no significant production of oestrogen from non-ovarian tissue, MCF-$7_{CA}$ cells serve as a local source of oestrogen to stimulate tumour formation in ovariectomized nude mice [74, 75] by aromatizing androstenedione. Thus inhibitors targeting aromatase and also antioestrogens that bind the ER can be studied in tumours formed from these cells. Therefore, the model has been employed to provide information that predicts the effects of these agents in the clinic and also as a guide to the development of new protocols to optimize their use in treatment. For example, the model has been used recently to investigate the effects of the non-steroidal aromatase inhibitors, letrozole and anastrozole and compared them with tamoxifen and Faslodex, the pure antioestrogen. As discussed below, combining aromatase inhibitors and antioestrogens was explored in the model. Inhibiting both oestrogen synthesis and oestrogen action simultaneously might be more effective than using either type of agent alone [79, 80].

In the xenograft model, tumours are developed by inoculating ovariectomized, female Balb/c mice (aged 4–6 weeks) with MCF-7 cells ($3 \times 10^7$ cells/ml in Matrigel) stably transfected with the human aromatase gene (MCF-$7_{CA}$). The cells were kindly provided by Dr. S. Chen (City of Hope, Duarte, CA, USA) [81]. As production of adrenal steroids in athymic mice is deficient [82], animals are injected subcutaneously from the day of inoculation throughout the experiment with 0.1 mg of androstenedione/mouse per day, the substrate for aromatization to oestrogens. Tumour growth is measured with calipers weekly and tumour volumes are calculated. When all tumours reach a measurable size (~300 mm$^3$), usually 28–35 days after inoculation, animals are assigned to groups with tumours of similar volume and treatment is begun. At autopsy, 4–6 h after the last injected dose, blood is collected and tumours are removed, cleaned, and weighed.

*Studies with anastrozole and letrozole*

We compared antioestrogens with aromatase inhibitors in the xenograft model to simulate first-line therapy in breast cancer patients. As anastrozole, letrozole, and exemestane are currently approved for use in the clinic, studies on the effects of these aromatase inhibitors are discussed below. We found that whereas the antioestrogens tamoxifen and faslodex, and the aromatase inhibitors letrozole and anastrozole, were highly effective in reducing tumour growth, both aromatase inhibitors were more effective than tamoxifen [79], as subsequently observed in clinical trials [31–35].

Treatment of mice with anastrozole (Arimidex, 5 μg/day), in contrast to tamoxifen (3 μg/day), caused significant inhibition of tumour growth compared to the controls ($P < 0.05$) [79]. Letrozole (10 μg/day) treatment was more potent than tamoxifen (60 μg/day) and fulvestrant (ICI 182,780; 5 mg/week) in controlling tumour growth, although both fulvestrant and letrozole showed regression of established tumours. Letrozole (5 μg/day) was also able to cause marked regression, even of large tumours.

The MCF-$7_{CA}$ tumours in the mouse model synthesize sufficient amounts of oestrogens to support oestrogen-dependent tumour growth and also to maintain the uterus of these ovariectomized animals at a weight similar to that of intact mice during metoestrus. Letrozole and anastrozole caused a decrease in the mean uterine weight compared to that of the control mice ($P < 0.01$). Neither of the aromatase inhibitors had oestrogenic effects on the uterus. The uterine weights of mice treated with tamoxifen were not significantly different from those of the control mice, consistent with previous reported findings reflecting the agonist/antagonists actions of tamoxifen [3, 46]. In contrast to tamoxifen, fulvestrant, considered to be a pure antioestrogen, blocked the actions of oestrogen on the uterus. Thus the uterine weights of fulvestrant-treated mice were similar to those treated with aromatase inhibitors. This indicates a difference in sensitivity of the effects of the two antioestrogens on the tumour and the uterus. Based on these results, it seems likely that aromatase inhibitors, even in long-term use, will not cause stimulation of the endometrium as reported in some women receiving tamoxifen. In recent clinical trials, no adverse effects on the endometrium have been observed in patients treated with the aromatase inhibitors letrozole, anastrozole, and exemestane [31, 34, 35].

*Sequential treatment with aromatase inhibitors and antioestrogens*

We observed previously that switching mice treated first with tamoxifen to second-line treatment with letrozole was effective in slowing tumour growth compared to continuing treatment with tamoxifen [80]. However, this strategy proved inferior to treatment with letrozole continued as first-line treatment. Unlike treatment with tamoxifen, tumours of mice treated with letrozole (10 μg/day) initially regressed. After extended treatment, tumours eventually grew during letrozole treatment. However, tumour-doubling time with letrozole was more than twice as long as with tamoxifen. Mice with tumours growing on letrozole treatment were then assigned to three groups so that each had similar mean tumour volumes at the start of second-line treatment. The groups were treated with either tamoxifen, a higher dose of letrozole (100 μg/day), or continued on letrozole (10 μg/day) treatment (Fig. 2) [80]. However, although the higher dose of letrozole slowed tumour growth, tumour volumes were not significantly different from those of groups treated with tamoxifen or continued on letrozole (10 μg/day). In another study, both antioestrogens, tamoxifen and fulvestrant, were ineffective as second-line therapy following letrozole treatment [83]. These results suggest that switching the animals from letrozole (10 μg/day) to antioestrogen treatment, is not beneficial for patients with tumours progressing on a therapeutically effective dose of letrozole.

*Combining treatment with aromatase inhibitors and tamoxifen*

As both antioestrogens and aromatase inhibitors are effective in treating breast cancer patients, combining these agents with different modes of action might result in greater anti-tumour efficacy than either alone. To study this hypothesis, low doses of the compounds were used which resulted in partial tumour suppression. Thus a greater reduction in tumour growth may be achieved by combining the two types of agent (Fig. 3). Since previous studies of 10 μg of letrozole/mouse per day caused almost complete regression of tumours, doses of 5 μg/day of letrozole and anastrozole were used in the combined treatments with tamoxifen at 3 μg/day. All compounds alone, or in combination at these doses, were effective in suppressing tumour growth in comparison to control mice. Weights of tumours removed at the end of treatment were significantly less for animals treated with the aromatase inhibitors letrozole and anastrozole than with tamoxifen ($P < 0.05$). However, treatment with either anastrozole or letrozole combined with tamoxifen did not produce greater reductions in tumour growth, as measured by tumour weight, than either aromatase inhibitor treatment alone, although tumour weights were reduced more than with tamoxifen alone [80, 84]. Oestrogen concentrations measured in tumour tissue of the letrozole-treated mice were markedly reduced from 460 to 20 pg/mg of tissue. Studies in patients treated with tamoxifen and letrozole suggest that the clearance rate of letrozole may be increased [85]. This could contribute to the combination being rather less effective than letrozole alone.

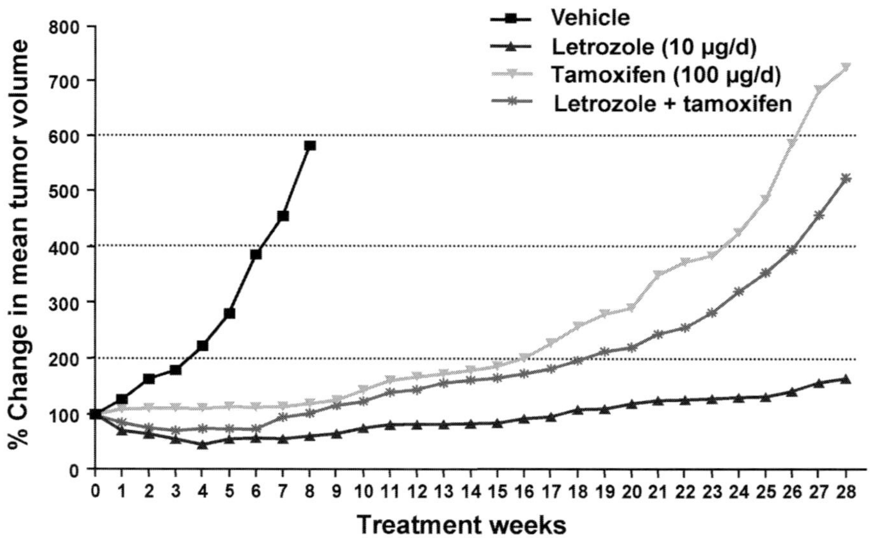


Figure 3. Effects of letrozole and tamoxifen and their combination on the growth of MCF-$7_{CA}$ breast tumour xenografts in female, ovariectomized, athymic nude mice. All mice received androstenedione (100 μg/day sc). Mice were divided into groups ($n = 20$ per group) and injected subcutaneously daily with vehicle, letrozole (10 μg/day) and/or tamoxifen (100 μg/day). Tumour volumes were measured weekly and are expressed as the percentage change relative to the initial tumour volume. Treatment with letrozole was statistically significantly better than the other treatments at 16 weeks. Tumour volumes were statistically significantly larger in the tamoxifen treatment group than in the letrozole treatment group at 28 weeks. Taken from [93].

These results suggest that combining non-steroidal aromatase inhibitors with tamoxifen does not improve treatment. Similar results were obtained when fulvestrant was combined with tamoxifen [84]. Tamoxifen may have a weak agonistic effect on the tumours which overrides the reduction in oestrogen concentrations by the aromatase inhibitors and which counteracts the effect of the pure antioestrogen. Subsequently, these results have been confirmed in the clinic by the Arimidex, Tamoxifen Alone or in Combination (ATAC) trial [31]. Patients with early breast cancer were treated with anastrozole, tamoxifen, or the combination. Treatment with anastrozole alone proved to be superior to tamoxifen, indicating for the first time that aromatase inhibitors were more effective in treating breast cancer patients than tamoxifen. However, treatment with the combination of anastrozole and tamoxifen was no better than with tamoxifen alone.

In recent studies, combining exemestane and tamoxifen showed that the combination was better than either tamoxifen or exemestane alone. This may reflect a dose-dependent effect by achieving a more complete oestrogen blockade [86].

*Loss of sensitivity with long-term letrozole*

The results of studies in the MCF-7 aromatase xenograft model indicate that although letrozole is useful in second-line therapy after tamoxifen [82], letrozole, as a single agent, was the most effective treatment and better alone than in combination with tamoxifen [80]. Nevertheless, following long-term tumour suppression during letrozole treatment, tumours eventually grew and were no longer sensitive either to the effects of the drug or to second-line treatment with the antioestrogens tamoxifen and fulvestrant [83].

Studies into the mechanisms involved in loss of response to letrozole treatment were carried out on tumours collected at several time points (weeks 4, 28, and 56) from mice during treatment with letrozole (10 pg/mouse per day; Fig. 4). Tumour extracts were analyzed for changes in protein expression using Western immunoblotting [87].

ER was increased after the first 4 weeks of letrozole treatment when tumours were regressing. After 56 weeks of letrozole treatment, tumours were growing and ER expression had decreased by 50% compared to control tumours. Interestingly, progesterone receptor expression was modestly increased despite low ER and suggests that ER activation could take place. Furthermore, phospho-ER (phosphorylated at Ser-67) was increased 2-fold in tumours collected at weeks 28 and 56, indicating that ligand-independent activation of ER may be occurring in tumours proliferating on letrozole. Expression of tyrosine kinase receptor erbB2 was increased throughout treatment with letrozole (weeks 4, 28, and 56). Also, phospho- (p-)Shc protein was increased by 2-fold at all time points with letrozole treatment, suggesting that the tumours may adapt to surviving without oestrogens by activating hormone-independent pathways. However, expression of the adapter protein Grb-2 was increased by 4-fold at weeks 28 and 56 in tumours that were actively growing

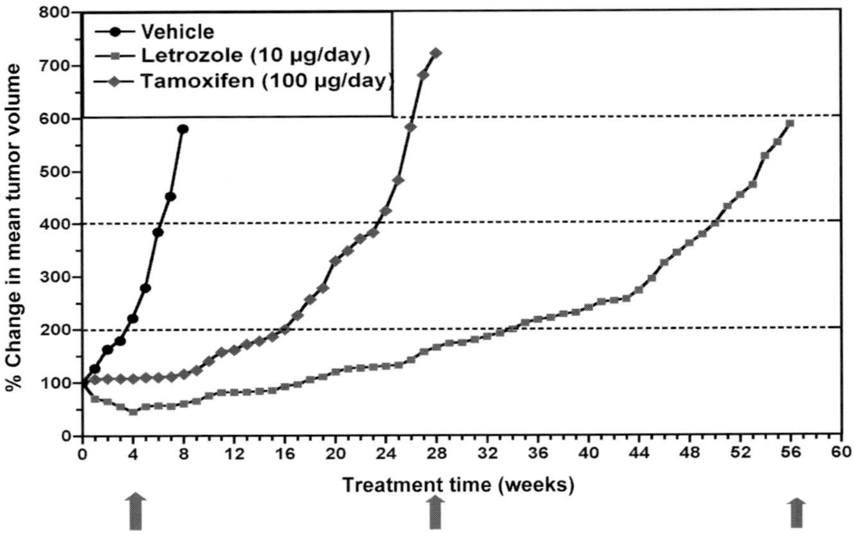


Figure 4. Effects of letrozole and tamoxifen on the growth of MCF-$7_{CA}$ breast tumour xenografts in female, ovariectomized, athymic, nude mice. All mice received androstenedione (100 µg/day sc). Mice were divided into groups ($n = 20$ per group) and injected subcutaneously daily with vehicle, letrozole (10 µg/day) or tamoxifen (100 µg/day). Tumour volumes were measured weekly and are expressed as the percentage change relative to the initial tumour volume. Treatment with vehicle, letrozole was statistically significantly better than the other treatments at 16 weeks. Tumour volumes were statistically significantly larger in the tamoxifen treatment group than in the letrozole treatment group at 28 weeks. Tumours were collected for analysis from some mice at weeks 4, 28, and 56 as indicated by the arrows.

on letrozole treatment. Phospho-mitogen-activated protein kinase (p-MAPK) was increased 2.3-fold in tumours that were responding to letrozole treatment at week 4 compared to vehicle-treated tumours, but was increased up to 6-fold in tumours growing on letrozole at weeks 28 and 56 (Fig. 5) [80]. These findings suggest that alternate signaling pathways are activated in tumours no longer sensitive to the effects of letrozole and support ER-mediated transcription despite the depletion of ligand (oestrogen) [87].

For further studies of the mechanisms involved in the loss of sensitivity to letrozole, tumour cells were isolated from the tumours of mice treated with letrozole for 56 weeks described above [80]. The cells were maintained in the presence of letrozole (1 nM) after isolation and designated long-term letrozole-treated, or LTLT, cells. The expression of signaling proteins in these cells was compared to the parental MCF-$7_{CA}$ cell line and also to a variant cell line derived from MCF-$7_{CA}$ by culturing the latter in steroid-depleted medium for 6 months (UMB-$1_{CA}$) [83]. In the latter cells, oestrogen deprivation resulted in a 2-fold increase in ER expression compared to the MCF-$7_{CA}$ cells. In contrast, ER expression was reduced in LTLT cells consistent with the decline in expression observed in the tumours of long-term letrozole-treated mice [87].

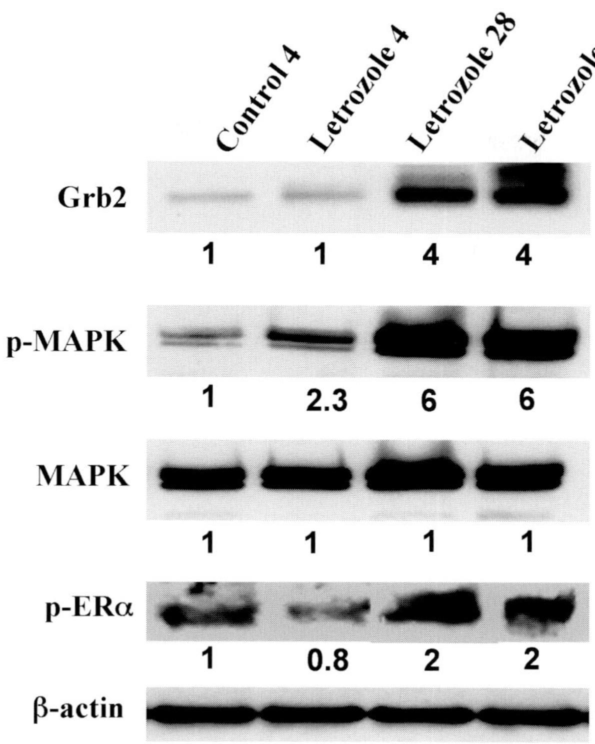


Figure 5. Expression of signaling proteins (p-ERα, Grb-2, MAPK, and p-MAPK) in tumour tissue from letrozole-treated mice at weeks 4, 28, and 56 compared to control tumours at 4 weeks. Protein extracts from tumour tissues were prepared by homogenizing the tissue and cells in lysis buffer. Proteins in the lysates were separated on a denaturing polyacrylamide gel and transferred to a nitrocellulose membrane. The protein-bound membranes were then incubated for 1 h at room temperature with 0.1% Tween 20 in PBS (PBS-T) and 5% non-fat dried milk to block non-specific binding to antibodies. The membranes were then incubated with respective primary antibodies in PBS-T milk for 1 h, and specific binding was visualized by using species-specific IgGs followed by enhanced chemiluminescence detection (ECL kit; Amersham Biosciences) and exposure to ECL X-ray film.

Interestingly, expression of erbB2 was increased in both cell lines compared to MCF-$7_{CA}$ cells as well as in letrozole-treated tumours, as indicated above. However, expression of adapter proteins (p-Shc and Grb-2) and signaling proteins p-MAPK, p-MEK1/2 (phospho-MAPK/extracellular-signal-regulated kinase kinase 1/2), p-Raf, p-p90 ribosomal S6 kinase, and pElk-1 were all increased in the LTLT cells but not in the UMB-$1_{CA}$ cells [88]. These results suggest that increase in Grb-2 expression in tumours proliferating on letrozole may be an important amplifier of the Ras-signaling pathway which leads to a further increase in activated MAPK and activation of ER in letrozole-treated cells. In contrast, UMB-$1_{CA}$ cells that had been deprived of oestrogen in their culture medium did not show increases in MAPK and associated signaling proteins. Instead, there was an increase in Akt and phosphoinositide 3-kinase activity in UMB-$1_{CA}$ cells compared to MCF-$7_{CA}$ cells [89]. Consistent with these results, cell proliferation of UMB-$1_{CA}$ could be inhibited by wortmannin and phosphoinositide 3-kinase inhibitors but not by MAPK inhibitors (PD98059) [89]. No increase in the Akt pathway was seen in the LTLT cells. This suggests that depriving the cells of oestrogen by aromatase inhibition results in activation of a different pathway from that of cells deprived of oestrogen in the medium. The activation of the Akt pathway in UMB-$1_{CA}$ cells appears to be similar to observations reported for MCF-7 cells deprived of oestrogen in culture [90] and involves crosstalk between the ER and Akt signaling [91]. UMB-$1_{CA}$ cells were susceptible to growth inhibition by the oestrogen downregulator Faslodex, whereas no such inhibition was apparent in the LTLT cells, consistent with results in the tumour model. Thus, activation of the Raf-MAPK pathway in LTLT cells may represent a more extreme form of oestrogen deprivation that could occur with long-term letrozole treatment. In the LTLT cells, proliferation was inhibited by the MAPK inhibitor PD98059 and the MEK1/2 inhibitor U0126 (obtained from Cell Signaling). These compounds were without effects on MCF-$7_{CA}$ proliferation. Iressa, an inhibitor of epidermal growth factor (EGF) tyrosine kinase, was effective in both UMB-$1_{CA}$ and LTLT cells, suggesting the involvement of EGF in the activation of this pathway [88].

*Combined treatment with letrozole and fulvestrant*

Evidence for the importance of ER in the activation of alternate signaling pathways was gained in a study combining the ER downregulator fulvestrant with letrozole [92]. Although the combination of the two non-steroidal inhibitors with tamoxifen had not shown improved results, we hypothesized that the combination of fulvestrant with letrozole may be more effective treatment than either compound alone. As the antioestrogen fulvestrant causes ER degradation, more complete oestrogen blockade may be achieved when it is combined with letrozole. To test this possibility, mice with established tumours were injected subcutaneously daily with vehicle (control), fulvestrant (1 mg/day), letrozole (10 μg/day), or letrozole plus fulvestrant at the same doses (Fig. 6) [92]. Tumours in the control group had doubled their initial volume after 3 weeks. All treatments were effective in suppressing tumour growth compared to the control group ($P < 0.001$). Tumours were static for the first 4 weeks of treatment with fulvestrant (1 mg/day) but then they began to proliferate and had doubled in volume after 10 weeks of treatment. By week 17, tumour volumes were significantly larger in the group treated with fulvestrant alone compared to the letrozole-treated group ($P < 0.001$). Tumour volumes were reduced by 40% over the first 8 weeks of treatment with letrozole but returned to their initial size by 17 weeks. After 21 weeks of treatment, tumours doubled in volume. The effect of letrozole (10 μg/day) on tumour growth in the MCF-7 aromatase xenograft model suggests that this aromatase inhibitor is better than the pure antioestrogen fulvestrant (1 mg/day) in controlling tumour growth and delaying the time of tumour progression. However, when the two

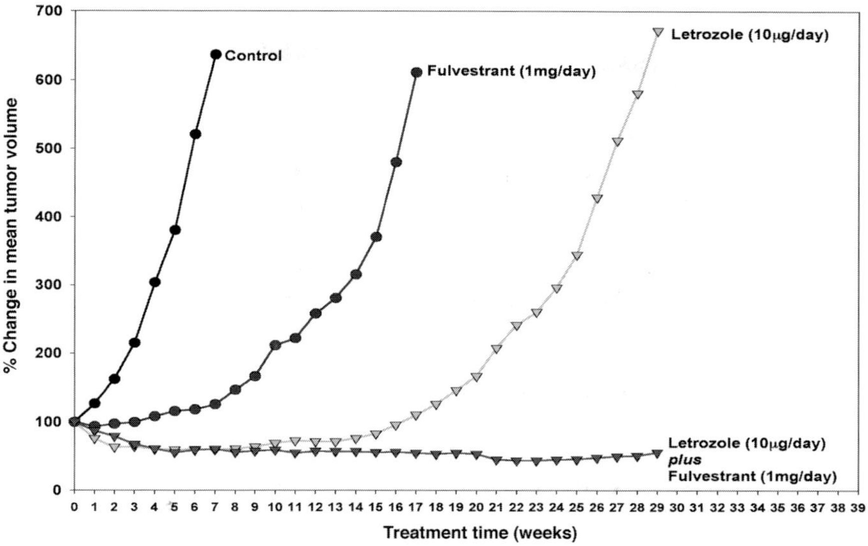


Figure 6. The effect of letrozole and fulvestant alone or in the combination on the growth of MCF-$7_{CA}$ breast tumour xenografts in female, ovariectomized, athymic, nude mice. All mice received androstenedione (100 μg/day sc). When tumours reached approximately 300 $mm^3$ animals were divided into four groups and injected subcutaneously daily with vehicle (control; $n = 6$), fulvestrant (1 mg/day; $n = 7$), letrozole (10 μg/day; $n = 18$), or letrozole (10 μg/day) plus fulvestrant (1 mg/day; $n = 5$). Tumour volumes were measured weekly and expressed as the percentage change in mean tumour volume relative to the initial size at week 0. At week 7, all treatments were significantly better at suppressing tumour growth compared to the control ($P < 0.0001$); all control animals were killed due to large tumour size. At week 17, letrozole was superior to fulvestrant in controlling tumour growth ($P < 0.001$). Also, treatment with letrozole plus fulvestrant was superior to fulvestrant alone ($P < 0.001$). Fulvestrant-treated mice were killed at week 17 due to large tumour size. At week 29, letrozole (10 μg/day) was less effective than letrozole plus fulvestrant in controlling tumour growth ($P = 0.0005$). Also, at week 29, tumour volume were statistically significantly larger in the letrozole treatment group, than in the combination ($P < 0.0001$).

drugs were combined, fulvestrant inhibiting oestrogen action and letrozole inhibiting oestrogen synthesis, tumour suppression was significantly greater than treatment with either letrozole or fulvestrant alone. This implies that some transcription via the ER may occur with fulvestrant treatment alone that is not completely blocked by the antioestrogen. The combined treatment resulted in tumour regression, which was maintained throughout the 29-week treatment period [92]. This result indicates that the combination of reducing oestrogen production and downregulating the ER could prevent or delay development of resistance to letrozole. Expression of erbB2 and MAPK were increased, relative to control samples, in tumours treated with letrozole and fulvestrant. However, there was no increase observed in tumours of mice treated with the combination [87]. These findings suggest that the combination of fulvestrant with letrozole could be more effective in breast cancer patients than these agents administered separately.

## Conclusion

In conclusion, a number of animal models have been utilized for preclinical studies of aromatase inhibitors. Some models are valuable for assessing the inhibition of oestrogen production from the ovary and also peripheral aromatization as well as on feedback regulation and other hormonal effects. Other models are relevant to anti-tumour activity. In the carcinogen-induced (DMBA/NMU) tumour models, the source of oestrogen is the ovary and, therefore, is under gonadotrophin feedback control. The advantages of the xenograft model are that tumours of human breast cancer cells are used and also that oestrogen is produced by aromatization from a non-ovarian source. These models come close to representing aspects of the situation in patients. However, there are no human breast cancer cell lines available at present in which growth is dependent on oestrogen and which express both ER and aromatase. Nevertheless, cells that are produced as a result of exposure to aromatase inhibitors are proving useful in culture and also as xenografts. These studies can provide valuable information for designing optimal treatment protocols to improve breast cancer treatment.

## References

1 Fornander T, Rutqvist LE, Cedermark B, Glass U, Matson A et al. (1989) Adjuvant tamoxifen in early breast cancer: occurrence of new primary cancer. *Lancet* 1: 117–120

2 Fornander T, Hellstrom AC, Moberger B (1993) Descriptive clinicopathologic study of 17 patients with endometrial cancer during or after adjuvant tamoxifen in early breast cancer. *J Natl Cancer Inst* 815: 1850–1855

3 Jordan VC, Rowsby L, Dix CJ, Prestwich G (1978) Dose-related effects of non-steroidal antiestrogens and estrogens on the measurement of cytoplasmic estrogen receptors in the rat and mouse uterus. *J Endocrinol* 78: 71–78

4 Schwarzel WC, Kruggel W, Brodie HJ (1973) Studies on the mechanism of estrogen biosynthesis. VII. The development of inhibitors of the enzyme system in the human placenta. *Endocrinology* 92: 866–880

5 Inkster SE, Brodie AMH (1989) Immunocytochemical studies of aromatase in early and full term human placental tissues: comparison with biochemical assays. *Biol Reprod* 41: 889–898

6 Fournet-Dulguerov N, MacLusky NJ, Leranth CZ, Todd R, Mendelson CR, Simpson ER, Naftolin F (1987) Immunohistochemical localization of aromatase cytochrome PF-450 and estradiol dehydrogenase in the syncytiotrophoblast of the human placenta. *J Clin Endocrinol Metab* 65: 757–764

7 Thompson EA Jr, Siiteri PK (1974) The involvement of human placental microsomal cytochrome P-450 in aromatization. *J Biol Chem* 249: 5373–5378

8 Brodie AMH, Schwarzel WC, Shaikh AA, Brodie HJ (1977) The effect of an aromatase inhibitor, 4-hydroxy-4-androstene-3,17-dione, on estrogen dependent processes in reproduction and breast cancer. *Endocrinology* 100: 1684–1694

9 Marsh DA, Brodie HJ, Garrett WM, Tsai-Morris CH, Brodie AMH (1985) Aromatase inhibitors-synthesis and biological activity of androstenedione derivatives. *J Med Chem* 28: 788–795

10 Brodie AMH, Garrett WM, Hendrickson JR, Marcotte PA, Robinson CH (1981) Inactivation of aromatase activity in placental and ovarian microsomes by 4-hydroxyandrostene-3,17-dione and 4-acetoxyandrostene-3,17-dione. *Steroids* 38: 693–702

11 Covey DF, Hood WF (1982) Aromatase enzyme catalysis is involved in the potent inhibition of estrogen biosynthesis caused by 4-acetoxy and 4-hydroxy-4-androstene-3,17-dione. *Mol Pharmacol* 21: 173–180

12 Covey DF, Hood WF, Parikh VD (1981) 10β-Propynyl-substituted steroids: mechanism-based enzyme-activated irreversible inhibitors of estrogen biosynthesis. *J Biol Chem* 256: 1076–1079
13 Marcotte PA, Robinson CH (1982) Synthesis and evaluation of 10β-substituted estrene-3,17-diones as inhibitors of human placental microsomal aromatase. *Steroids* 39: 325–344
14 Metcalf BW, Wright CL, Burkhart JP, Johnston JO (1981) Substrate-induced inactivation of aromatase by allenic and acetylenic steroids. *J Am Chem Soc* 103: 3221–3222
15 Bellino FL, Gilani SSH, Eng SS, Osawa Y, Duax WL (1976) Active-site inactivation of aromatase from human placental microsomes by brominated androgen derivatives. *Biochemistry* 15: 4730–4736
16 Bruegemeier RW, Floyd EE, Counsell RE (1978) Synthesis and biochemical evaluation of inhibitors of estrogen biosynthesis. *J Med Chem* 21: 1007–1011
17 Bruegemeier RW, Li PK, Snider CE, Darby MV, Katlic NE (1987) 7α-Substituted androstenediones as effective *in vitro* and *in vivo* inhibitors of aromatase. *Steroids* 50: 167–178
18 Bruegemeier RW, Li PK, Chen HH, Moh PP, Katlic NE (1990) Biochemical pharmacology of new 7-substituted androstenediones as inhibitors of aromatase. *J Steroid Biochem Mol Biol* 37: 379–385
19 Zaccheo T, Giudici D, Ornati G, Panzeri A, Di Salle E (1991) Comparison of the effects of the irreversible aromatase inhibitor exemestane with atamestane and MDL-18962 in rats with DMBA-induced mammary tumours. *Eur J Cancer* 27: 1145–1150
20 Giudici D, Ornati G, Briatico G, Buzzetti F, Lombardi P, Salle ED (1988) 6-Methylenandrosta-1,4-diene-3,17-dione (FCE 24304): A new irreversible aromatase inhibitor. *J Steroid Biochem Mol Biol* 30: 391–394
21 Thompson EA Jr, Siiteri PK (1974) Utilization of oxygen and reduced nicotinamide adenine dinucleotide phosphate by human placental microsomes during aromatization of androstenedione. *J Biol Chem* 249: 5364–5372
22 Siiteri PK, Thompson EA (1975) Studies of human placental aromatase. *J Steriod Biochem* 6: 317–322
23 Cash R, Brough AJ, Cohen MNP, Satoh PS (1967) Aminoglutethimide (Elipten-CIBA) as an inhibitor of adrenal steroidogenesis. Mechanism of action and therapeutic trial. *J Clin Endocrinol Metab* 27: 1239–1248
24 Santen RJ, Santner S, Davis B, Velhuis J, Samojlik E, Ruby E (1978) Aminoglutethimide inhibits extraglandular estrogen production in postmenopausal women with breast carcinoma. *J Clin Endocrinol Metab* 47: 1257–1265
25 Vytautas UI, Whipple CA, Salhanick HA (1977) Steroselective inhibition of cholesterol side chain cleavage by enantiomers of aminoglutethimide. *Endocrinology* 101: 89–92
26 Brodie AMH, Longcope C (1980) Inhibition of peripheral aromatization by aromatase inhibitors, 4-hydroxy- and 4-acetoxy-androstene-3,17-dione. *Endocrinology* 106: 19–21
27 Brodie AMH, Garrett WM, Hendrickson JR, Tsai-Morris CH (1982) Effects of 4-hydroxyandrostenedione and other compounds in the DMBA breast carcinoma model. *Cancer Res* 42: 3360s–3364s
28 Goss PE, Coombes RL, Powles TJ, Dowsett M, Brodie AMH (1986) Treatment of advanced postmenopausal breast cancer with aromatase inhibitor, 4-hydroxyandrostenedione-phase 2 report. *Cancer Res* 46: 4823–4826
29 Coombes CR, Goss P, Dowsett M, Gazet JC, Brodie A (1984) 4-Hydroxyandostenedione in treatment of postmenopausal patients with advanced breast cancer. *Lancet* 2: 1237–1239
30 Paridaens R, Dirix L, Lohrisch C, Beex L, Nooij M, Cameron D et al. (2003) Mature results of a randomized phase II multicenter study of exemestane *versus* tamoxifen as first-line hormone therapy for postmenopausal women with metastatic breast cancer. *Ann Oncol* 14: 1391–1398
31 Baum M, Buzdar A, Cuzick J, Forbes J, Houghton J, Klijn IG et al. (2003) Anastrozole alone or in combination with tamoxifen *versus* tamoxifen alone for adjuvant treatment of postmenopausal women with early-stage breast cancer: results of the ATAC (Arimidex, Tamoxifen Alone or in Combination) trial efficacy and safety update analyses. *Cancer* 98: 1802–1810
32 Mouridsen H, Gershanovich M, Sun Y, Perez-Carrion R, Boni C, Monnier A et al. (2003) Phase III study of letrozole *versus* tamoxifen as first-line therapy of advanced breast cancer in postmenopausal women: analysis of survival and update of efficacy from the International Letrozole Breast Cancer Group. *J Clin Oncol* 21: 2101–2109
33 Nabholtz JM, Buzdar A, Pollak M, Harwin W, Burton G, Mangalik A et al. (2000) Anastrazole is superior to tamoxifen as first-line therapy for advanced breast cancer in postmenopausal women:

results of a North American multicenter randomized trial. Arimidex Study Group. *J Clin Oncol* 18: 3758–3767

34 Coombes RC, Hall E, Gibson LJA (2004) Randomized trial of exemestane after two to three years of tamoxifen therapy in postmenopausal women with primary breast cancer. *N Eng J Med* 350: 1081–1092

35 Goss PE, Ingle JN, Martino S, Robert NJ, Muss HB, Piccart MJ et al. (2003) A randomized trial of letrozole in postmenopausal women after five years of tamoxifen therapy for early-stage breast cancer. *New Engl J Med* 349: 1793–1802

36 Brodie AMH, Marsh DA, Wu JT, Brodie HJ (1979) Aromatase inhibitors and their use in controlling estrogen dependent processes. *J Steroid Biochem* 11: 107–112

37 Brodie AMH, Wu JT, Marsh DA, Brodie HJ (1978) Aromatase inhibitors III. Studies on the antifertility effects of 4-acetoxy-4-androstene-3,17-dione. *Biol Reprod* 18: 365–370

38 Fisher CR, Graves KH, Parlow AF, Simpson ER (1998) Characterization of mice deficient in aromatase (ArKO) because of targeted disruption of the CYP19 gene. *Proc Natl Acad Sci USA* 95: 6965–6970

39 Couse JF, Korach KS (1999) Estrogen receptor null mice: what have we learned and where will they lead us? *Endocr Rev* 20: 358–417

40 Tekmal R, Ramachandra N, Gubba S, Durgam VR, Manitone J, Toda K et al. (1996) Overexpression of int-5/aromatase in mammary glands of transgenic mice results in the induction of hyperplasia and nuclear abnormalities. *Cancer Res* 56: 3180–3185

41 Gill K, Keshava N, Manitone J, Tekmal RR (2001) Overexpression of aromatase in transgenic male mice results in the induction of gynecomastia and other biochemical changes in mammary glands. *J Steroid Biochem Mol Biol* 77: 13–18

42 Fowler KA, Gill K, Kirma N, Dillehay DL, Tekmal RR (2000) Overexpression of aromatase leads to development of testicular leydig cell tumors: an *in vivo* model for hormone-mediated testicular cancer. *Am J Pathol* 156: 347–353

43 Keshava N, Fang C, Bhalla KN, Tekmal RR (1995) Environmental mutagen (DMBA) acts synergistically in int-5/aromatase transgenic mice that have mammary estrogen activity. *Mutat Res* 379: S151

44 Brodie AMH, Marsh DA, Brodie HJ (1979) Aromatase inhibitors IV. Regression of estrogen-dependent mammary tumors in the rat with 4-acetoxy-4-androstene-3,17-dione. *J Steroid Biochem* 10: 423–429

45 Brodie AMH, Garrett W, Hendrickson JM, Marsh DA, Brodie HJ (1982) The effect of 1,4,6-androstatriene-3,17-dione (ATD) on DMBA-induced mammary tumors in the rat and its mechanism of action *in vivo*. *Biochem Pharmacol* 31: 2017–2023

46 Jordan VC (1987) Lab models of breast cancer to aid the elucidation of antiestrogenation. *J Lab Clin Med* 106: 267–277

47 Jordan VC (1982) Lab models of hormone-dependent cancer. In: Furr BJA (ed.): *Clinics in oncology*. WB Saunders Co, London, 21–40

48 Huggins C, Briziarelli G, Sutton H (1959) Rapid induction of mammary carcinoma in the rat and the influence of hormones on tumors. *J Exp Med* 109: 25–42

49 DeSombre ER, Arbogast LY (1974) Effect of the antiestrogen C1628 on the growth of rat mammary tumors. *Cancer Res* 34: 1971–1976

50 Thompson EA, Hemsell D, McDonald PC, Siiteri PK (1974) Inhibition of aromatization by steroidal drugs. *J Biol Chem* 5: 315

51 Uzgiris VI, Graves P, Salhanick HA (1977) Liquid modification of corpus luteum mitochondrial cytochrome P-450 spectra and cholesterol monooxygenation: an assay of enzyme-specific inhibitors. *Endocrinology* 101: 89–92

52 Sherman BM, Chapler FK, Crickard K, Wycoff D (1979) Endocrine consequences of continuous antiestrogen therapy with tamoxifen in premenopausal women. *J Clin Invest* 64: 398–404

53 Samojilik E, Veldhuis JD, Wells SA, Santen RJ (1980) Preservation of androgen secretion during estrogen suppression with aminoglutethimide in the treatment of metastatic breast carcinoma. *J Clin Invest* 65: 602–612

54 Wing LY, Garrett WM, Brodie MH (1985) The effect of aromatase inhibitors, aminogluthimide and 4-hydroxyandrostenedione on cyclic rats with DMBA-induced mammary tumors. *Cancer Res* 45: 2425–2428

55 Hemsell DL, Gordon J, Breuner PF, Siiteri PK, MacDonald PC (1974) Plasma precursors of estrogen. II. Correlation of extent of conversion of plasma androstenedione to estrone with age. *J Clin*

*Endocrinol Metab* 38: 476–479
56 Thorsen T, Tangen M, Stoa KF (1982) Concentrations of endogeneous estradiol as related to estradiol receptor sites in breast tumor cytosol. *Eur J Cancer Clin Oncol* 18: 333–337
57 van Landeghem AAJ, Portman J, Nabauurs M (1985) Endogeneous concentration and subcellular distribution of estrogens in normal and malignant human breast tissue. *Cancer Res* 45: 2900–2906
58 Blankenstein MA, Maitimu-Smeele I, Donker GH, Daroszewski J, Milewicz A, Thijssen JHH (1992) On the significance of *in situ* production of oestrogens in human breast cancer tissue. *J Steroid Biochem Mol Biol* 41: 891–896
59 Perel E, Blackstein ME, Killinger DW (1982) Aromatase in human breast carcinoma. *Cancer Res* (suppl.) 42: 3369s–3372s
60 James VHT, McNeill JM, Lai L, Newton CJ, Ghilchik MW, Reed MJ (1987) Aromatase activity in normal breast and breast tumor tissue: *in vivo* and *in vitro* studies. *Steroids* 50: 269–279
61 Killinger DW, Perel E, Daniilescu D, Kharlip L, Blackstein ME (1987) Aromatase activity in the breast and other peripheral tissues and its therapeutic regulation. *Steroids* 50: 523–535
62 Miller WR, Hawkins RA, Forrest APM (1982) Significance of aromatase activity in human breast cancer. *Cancer Res* 42 (suppl.): 3365s–3368s
63 Peice T, Aitken J, Head J, Mahendroo M, Means G, Simpson E (1992) Determination of aromatase cytochrome P-450 messenger ribonucleic acid in human breast tissue by competitive polymerase chain reaction amplification. *J Clin Endocrinol Metab* 174: 1247–1252
64 Koos RD, Banks PK, Inkster SE, Yue W, Brodie AMH (1993) Detection of aromatase and keratinocyte growth factor expression in breast tumors using reverse transcription-polymerase chain reaction. *J Steroid Biochem Mol Biol* 45: 217–225
65 Lu Q, Nakamura J, Savinov A, Yue W, Weisz J, Dabbs DJ et al. (1996) Expression of aromatase protein and mRNA in tumor epithelial cells and evidence of functional significance of locally produced estrogen in human breast cancer. *Endocrinology* 137: 3061–3068
66 Lipton A, Santen RJ, Santen SJ, Harvey HA, Peil PD, White-Hershey D et al. (1987) Aromatase activity in primary and metastatic human breast cancer, *Cancer* 59: 779–783
67 Brodie AMH, Lu Q, Long BJ, Fulton A, Chen T, MacPherson N et al. (2001) Aromatase and COX-2 expression in human breast cancers. *J Steroid Biochem Mol Biol* 1607: 1–7
68 Brueggemeier RW, Quinn AL, Parrett ML, Joarder FS, Harris RE, Robertson FM (1999) Correlation of aromatase and cyclooxygenase gene expression in human breast cancer specimens. *Cancer Lett* 140: 27–35
69 Simpson ER, Mehendroo MS, Means GD, Kilgore MW, Corbin CJ, Mendelson CR (1993) Tissue-specific promoters regulate cytochrome P-450 expression. *J Steroid Biochem Mol Biol* 44: 321–330
70 Simpson ER, Ackerman GE, Smith ME, Mendelson CR (1981) Estrogen formation in stromal cells of adipose tissue of women: Induction of glucocorticosteroids. *Proc Natl Acad Sci USA* 78: 5690–5694
71 Johnston JO, Wright CL, Schumaker RC (1989) Human trophoblast xenografts in athymic mice: a model for peripheral aromatization. *J Steroid Biochem Mol Biol* 33: 521–529
72 Franz C, Longcope C (1979) Androgen and estrogen metabolism in male Rhesus monkeys. *Endocrinology* 105: 869–874
73 Lonning PE, Geisler J, Bhatnagar A (2003) Development of aromatase inhibitors and their pharmacological profile. *Am J Clin Oncol* 4:S3–S8
74 Yue W, Zhou D, Chen S, Brodie A (1994) A new nude mouse model for postmenopausal breast cancer using MCF-7 cells transfected with the human aromatase gene. *Cancer Res* 54: 5092–5095
75 Yue W, Wang J, Savinov A, Brodie A (1995) The effect of aromatase inhibitors on growth of mammary tumors in a nude mouse model. *Cancer Res* 55: 3073–3077
76 Mattern J, Bak M, Hahn EW, Volm M (1988) Human tumor xenografts as model for drug testing. *Cancer Metas Rev* 7: 263–284
77 Osborne CK, Jarman M, McCague R, Coronado EB, Hilsenbeck SG, Wakeling AE (1994) The importance of tamoxifen metabolism in tamoxifen-stimulated breast tumor growth. *Cancer Chemother Pharm* 34: 89–95
78 Kinoshita Y, Chen S (2003) Induction of aromatase expression in breast cancer cells through nongenomic action of ER$\alpha$. *Cancer Res* 63: 3546–3555
79 Lu Q, Yue W, Wang J, Liu Y, Long BJ, Brodie A (1998) The effect of aromatase inhibitors and antiestrogens in the nude mouse model. *Breast Cancer Res Treat* 50: 63–71
80 Long BJ, Jelovac D, Handratta V, Thiantanawat A, MacPherson N, Ragaz J, Brodie AM (2004)

Therapeutic strategies using the aromatase inhibitor letrozole and tamoxifen in breast cancer model. *J Natl Cancer Inst* 96: 456–465

81 Zhou D, Pompon D, Chen S (1990) Stable expression of human aromatase complementary DNA in mammalian cells: a useful system for aromatase inhibitor screening, *Cancer Res* 50: 6949–6954

82 Rebar RW, Morandini IC, Erickson GF, Petze JE (1981) The hormonal basic of reproductive defects in athymic mice: diminished gonadotropin concentration in prepubertal females. *Endocrinology* 108: 120–126

83 Long BJ, Jelovac D, Thiantanawat A, Brodie AM (2002) The effect of second-line antiestrogen therapy on breast tumor growth after first-line treatment with the aromatase inhibitor letrozole: long-term studies using the intratumoral aromatase postmenopausal breast cancer model. *Clin Cancer Res* 8: 2378–2388

84 Lu Q, Liu Y, Long BJ, Grigoryev D, Gimbel M, Brodie A (1999) The effect of combining aromatase inhibitors with antiestrogens on tumor growth in a nude mouse model for breast cancer. *Breast Cancer Res Treat* 57: 183–192

85 Dowsett M, Pfister C, Johnston SR, Miles DW, Houston SJ, Verbeek JA et al. (1999) Impact of tamoxifen on the pharmacokinetics and endocrine effects of the aromatase inhibitor letrozole in postmenopausal women with breast cancer. *Clin Cancer Res* 5: 2338–2343

86 Jelovac D, Macedo L, Handratta V, Long BJ, Goloubeva OG, Brodie AMH (2004) Preclinical studies evaluating the anti-tumor effects of exemestane alone or combined with tamoxifen in a postmenopausal breast cancer model. *Clin Cancer Res* 10: 7375–7381

87 Jelovac D, Sabnis G, Long BJ, Goloubeva OG, Brodie AMH (2005) Activation of MAPK in xenografts and cells during prolonged treatment with aromatase inhibitor letrozole. *Cancer Res* 65: 5380–5389

88 Brodie A, Jelovac D, Long B, Macedo, L, Goloubeva O (2005) Model systems: mechanisms involved in the loss of sensitivity to letrozole. *J Steroid Biochem Mol Biol* 95: 41–48

89 Sabnis GJ, Jelovac D, Long B, Brodie A (2005) The role of growth factor receptor pathways in human breast cancer cells adapted to long term estrogen deprivation. *Cancer Res* 65: 3903–3910

90 Yue W, Wang JR, Conaway MR, Li Y, Santen RS (2003) Adaptive hypersensitivity following long-term estrogen deprivation: Involvement of multiple signal pathways. *J Steroid Biochem Mol Biol* 86: 265–274

91 Shou J, Massarweh S, Osborne CK, Wakeling AE, Au S, Weiss H, Schiff P (2004) Mechanisms of tamoxifen resistances Increased estrogen receptor-HEP2/neu cross-talk in EP/HEP2-positive breast cancer. *J Natl Cancer Inst* 96: 926–935

92 Jelovac D, Macedo L, Goloubeva OG, Handratta V, Brodie AMH (2005) Additive antitumor effect of aromatase inhibitor letrozole and antiestrogen fulvestrant in a postmenopausal breast cancer model. *Cancer Res* 65: 5439–5444

Aromatase Inhibitors
Edited by B.J.A. Furr



# Clinical pharmacology of aromatase inhibitors

Jürgen Geisler and Per Eystein Lønning

*Department of Medicine, Section of Oncology, Haukeland University Hospital, 5021 Bergen, Norway*

## Introduction

The preclinical pharmacology of aromatase inhibitors has been reviewed in chapter by W.R. Miller. However, preclinical pharmacology may not always be directly extrapolated to the *in vivo* setting in humans. Thus a drug effect in the human body will depend on factors in addition to its effect on the target enzyme, like general pharmacokinetics, tissue penetration and cellular uptake. In contrast to the selective oestrogen receptor modulators (SERMs), for which a simple biochemical parameter *in vivo* is lacking, aromatase inhibitors may be assessed by their ability to modulate their target, the aromatase enzyme. Whereas we now have data available comparing the effect of different compounds on total body aromatase inhibition, our understanding of the effects of these compounds at the tissue level, in particular with respect to local effects in the normal breast as well as breast tumour tissue, is still incomplete.

A key point understanding the pharmacology of any compound *in vivo* relates to its pharmacokinetics. A detailed description of the disposition of the different compounds is beyond the scope of this chapter but the reader is referred to a recent, more comprehensive review on the subject [1]; a brief description of the pharmacokinetics of the three third-generation inhibitors will, however, be provided.

## Clinical pharmacokinetics

Different methods are available for measuring plasma levels of anastrozole as well as letrozole and exemestane [2–5]. However, so far, no study has reported tissue levels of any of these compounds in humans.

Considering absorption, letrozole is the only compound for which this has been assessed in humans by comparing parenteral and orally administered compound [6]. Whereas in-house animal experiments suggest anastrozole may be well absorbed [7], and absorption of exemestane has been partly evaluated through oral administration of radioactive compounds [8], the precise bioavailability of these two compounds in humans is unknown.

Anastrozole and letrozole both seem to be associated with a terminal plasma half-life of about 48 h following administration of a single dose [2, 6, 9]. In contrast, the half-life of exemestane is probably about 24 h [10]. Whereas no study has determined tissue drug concentrations, it is important to recognize that following termination of treatment with anastrozole or letrozole plasma oestrogen levels may need up to 4 weeks to recover [11]. However, it is difficult from these data to extrapolate about the exact tissue half-life of the compounds. Due to their potency, it is likely that these drugs, even at low concentrations, may express some influence on the aromatase enzyme. Full recovery of tissue oestrogen levels may also depend on time delay to achieve equilibrium with oestrone sulphate, which is known to have a longer terminal half-life compared to unconjugated oestrogens in plasma [12].

An issue of particular interest has been potential drug interactions between aromatase inhibitors and other compounds. Whereas the first-generation aromatase inhibitor, the barbiturate aminoglutethimide, was known to be a potent inducer of mixed-function oxidases [13], including enhancing the metabolism of tamoxifen as well as progestins [14–16], anastrozole and letrozole were both found to have no effect on total body disposition of tamoxifen [17, 18]. Interestingly, tamoxifen treatment was found to suppress modestly plasma concentration of both anastrozole and letrozole [17, 19]. This 30–40% reduction of plasma levels of anastrozole and letrozole is not expected to impair plasma oestrogen suppression. Considering exemestane, no drug interaction has been reported so far.

## *In vitro* evaluation of aromatase inhibitors and inactivators

*In vitro* assessment of aromatase inhibitors and inactivators is in general conducted using placental or ovarian aromatase as a test substrate [20]. The results of *in vitro* evaluations of aromatase inhibitors have been reviewed by others [21–23]. Table 1 summarizes the *in vitro* findings and gives references to the original publications [24–30]. Whereas *in vitro* data may underpin the potency of individual drugs, suggesting which one is to be chosen for clinical development and testing, the importance of *in vivo* assessment of endocrine effects is illustrated by the comparison of fadrozole and letrozole. Thus, whereas fadrozole was revealed to be more potent than letrozole *in vitro* [31, 32], letrozole was superior *in vivo* [33, 34]. Whether the discrepancy between *in vitro* and *in vivo* findings is related to differences in the pharmacokinetic disposition alone or other factors is not known [6, 9, 35].

## Effects on *in vivo* aromatization and plasma oestrogen levels

Since the pioneering study of Santen et al. [36] using double-radioisotope tracer techniques to show aminoglutethimide inhibited the conversion of

Table 1. *In vitro* potency of aromatase inhibitors

| Compound | $IC_{50}$ arom (nM) | Reference |
|---|---|---|
| Aminoglutethimide | 1900 | [24] |
| Fadrozole | 5 | [25] |
| Vorozole | 1.3 | [26] |
| Anastrozole | 15 | [27] |
| Letrozole | 11.5 | [28] |
| 4-Hydroxyandrostenedione | 62 | [29] |
| Exemestane | 30 | [30] |

$IC_{50}$ arom means the drug concentration causing 50% aromatase inhibition in a given test system, using human placental aromatase.

androstenedione to oestrone *in vivo*, such tracer studies have been considered the 'gold standard' in assessing the efficacy of aromatase inhibitors *in vivo.* The main reason for this has been difficulties creating plasma oestrogen assays with sufficient sensitivity to define the full extent of oestrogen suppression achieved with these potent novel third-generation compounds [37]. Assessment of total body aromatization *in vivo* with tracer techniques may be achieved by one of two methods. The first includes infusing $^{3}H$-labelled androstenedione and $^{14}C$-labelled oestrone to achieve plasma steady-state levels, after which the isotope fraction in plasma oestrone is measured. In the second method, the same tracer compounds are given by bolus injection, followed by collection of urine for 4 days with measurement of the isotope ratio in the oestrogen metabolites. The latter method [38] has proved to be the most sensitive, enabling detection of more than 99% aromatase inhibition in the majority of patients [39]. Using this method, we were able to classify aromatase inhibitors in the first- and second-generation compounds, which may achieve up to 85–90% aromatase inhibition *in vivo* [33, 40–42], and the recent third-generation drugs, which all cause ≥98% aromatase inhibition (Tab. 2) [34, 39, 43, 44]. Moreover, applying recent sensitive assays for plasma oestrogen determinations, we were able to detect suppression of plasma oestrone sulphate, closely corroborating the degree of aromatase inhibition [37].

Whereas there have been dissenting opinions about whether total body aromatization and plasma oestrogen levels reflect what is happening in tissues, interestingly these biochemical findings are consistent with clinical observations. Thus randomized studies comparing the first- and second-generation compounds aminoglutethimide, formestane and fadrozole, either to megestrol acetate or tamoxifen in metastatic disease, reveal no clinical superiority for any of these compounds compared to conventional treatment [45–49]. In contrast, as will be reviewed elsewhere in this volume, the novel, more potent, third-generation compounds provided clinical superiority.

Table 2. Effects of different aromatase inhibitors and inactivators on whole-body aromatisation

| Compound | Drug dose (mg) | Aromatase inhibition (%) | Reference |
|---|---|---|---|
| Aminoglutethimide | 250 qid | 90.6 | [40] |
| Formestane (*per os*) | 125 od, 125 bid, 250 od | 72.3, 70, 57.3 | [41] |
| Formestane (intramuscularly) | 250 2w, 500 2w, 500 w | 84.8, 91.9, 92.5 | [42] |
| Rogletimide | 200 bid, 400 bid, 800 bid | 50.6, 63.5, 73.8 | [40] |
| Fadrozole | 1 bid, 2 bid | 82.4, 92.6 | [33] |
| Anastrozole | 1 od, 10 od | 96.7, 98.1 | [43] |
| | 1 od | 97.3 | [34] |
| Letrozole | 0.5 od, 2.5 od | 98.4, 98.9 | [39] |
| | 2.5 od | >99.1 | [34] |
| Exemestane | 25 od | 97.9 | [44] |

All values were determined by the same assay at the Academic Department of Biochemistry, Royal Marsden Hospital, London, UK (head: Professor M. Dowsett) and the Breast Cancer Research Group at the Haukeland University Hospital in Bergen, Norway (head: Professor P.E. Lønning). Abbreviations: od, once daily; bid, twice daily; qid, four times daily; w, weekly; 2w, every 2 weeks.

## Breast cancer tissue oestrogen levels

The problems mentioned above with respect to sensitive assays for plasma oestrogen levels relate to tissue oestrogen levels as well. Assessment of tissue oestrogen levels in general, but in particular during treatment with aromatase inhibitors, requires assays with a high sensitivity and specificity, usually involving several purification steps (like HPLC) followed by radioimmunoassay [50]. Interesting differences between plasma and tissue oestrogen levels may be observed when looking at the ratios between the different oestrogen fractions. For example, whereas oestrone sulphate is the dominant oestrogen fraction in the circulation of postmenopausal women, showing a concentration about 10–20-fold the concentrations of oestrone and oestradiol respectively [51, 52], the dominant oestrogen in the tissue, in particular in oestrogen receptor-/progesterone receptor-positive breast tumours, is oestradiol. In oestrogen receptor-positive breast cancer samples from postmenopausal women, the concentration of oestradiol is about 10-fold the concentration measured in the plasma. In contrast to others [53], we found breast cancer tissue oestrone sulphate levels to be much lower compared to plasma oestrone sulphate levels [51, 54].

The observation that tissue levels of oestrone and oestradiol are higher compared to plasma levels is consistent with current knowledge concerning disposition of oestrogens in postmenopausal women. Oestrogens are synthesized in most peripheral tissues (see [23] for references) from circulating androgens, mainly androstenedione, secreted by the adrenal gland and, to a minor extent, probably the postmenopausal ovary [55]. Thus we believe that the concentration gradient between tissue and plasma is due to passive diffusion, as circulating oestrogens arise by leakage from the tissue following metabolism and excretion by the liver and kidney, respectively [56]. Accordingly, the assessment of total body aromatization with use of tracer techniques estimates the sum of oestrogens produced in the peripheral tissues and should be considered as a surrogate marker for non-glandular oestrogen production.

A different issue relates to local oestrogen synthesis within the tumour tissue. Interestingly, there is a substantial variation in oestrogen levels between different tumours. This probably reflects differences regarding expression of the aromatase enzyme, although differences with respect to local oestrogen metabolism may be relevant as well [57, 58]. Whereas only one aromatase gene has been identified, this contains at least 10 different promoters [59]. The promoters II, I.3 and I.7 are particularly active in breast cancer tissue [59]. Notably, these promoter regions are stimulated by different growth factors and interleukins known to be synthesized in breast tumours, probably contributing to the high local oestrogen concentrations observed in some human breast tumours [54]. It is noteworthy that tissue oestrogen concentrations seem to be much higher in oestrogen receptor-positive compared to -negative tumours [52]. Beside aromatase, several other enzyme systems (see [51] for references) are involved in oestrogen synthesis and conversion in postmenopausal women, such as steroid sulphatase, oestrogen sulphotransferase and 17β-hydroxysteroid dehydrogenase type 1 and 2.

Whereas the influence of aromatase inhibitors on tissue oestrogen levels has been evaluated in several studies [54, 60–62], each study involved a limited number of patients only. An overview has recently been published [51]. Concerning the third-generation aromatase inhibitors, significantly decreased tissue oestrogen levels in breast tissue samples have been found during treatment with anastrozole [54] and letrozole [62]. Data about the influence of exemestane on tissue oestrogen levels are currently not available.

## Summary

Third generation aromatase inhibitors (anastrozole, letrozole and exemestane) differ to previous compounds with respect to their biochemical efficacy. While in general there is a good consistency between *in vitro* and *in vivo* effects, notable there may be important differences, as illustrated by comparing fadrozole and letrozole. This is due to the fact that *in vivo* effects also depend on local tissue and total body drug disposition. Whether the lack of cross-resistance between non-steroidal and steroidal compounds [11] may be explained by differential effects on the aromatase enzyme (enzyme inactivation versus enzyme inhibition) or by other factors, like slight androgen side-effects of the steroidal compounds [63], remains an open question.

## References


1 Lønning PE (2003) Clinical pharmacokinetics of aromatase inhibitors and inactivators. *Clin Pharmacokinet* 42: 619–631
2 Yuan J, Wang PQ, Ge SR et al. (2001) Pharmacokinetics of anastrozole in Chinese male volunteers. *Acta Pharmacol Sin* 22: 573–576
3 Pfister CU, Duval M, Godbillon J et al. (1994) Development, application and comparison of an enzyme immunoassay and a high-performance liquid chromatography method for the determination of the aromatase inhibitor CGS 20,267 in biological fluids. *J Pharm Sci* 83: 520–524
4 Marfil F, Pineau V, Sioufi A, Godbillon SJ (1996) High-performance liquid chromatography of the aromatase inhibitor, letrozole, and its metabolite in biological fluids with automated liquid-solid extraction and fluorescence detection. *J Chromatogr B Biomed Appl* 683: 251–258
5 Breda M, Pianezzola E, Benedetti MS (1993) Determination of exemestane, a new aromatase inhibitor, in plasma by high-performance liquid chromatography with ultraviolet detection. *J Chromatogr* 620: 225–231
6 Sioufi A, Gauducheau N, Pineau V et al. (1997) Absolute bioavailability of letrozole in healthy postmenopausal women. *Biopharm Drug Dispos* 18: 779–789
7 Plourde PV, Dyroff M, Dukes M (1994) Arimidex: a potent and selective fourth-generation aromatase inhibitor. *Breast Cancer Res Treat* 30: 103–111
8 Cocchiara G, Allievi C, Berardi A et al. (1994) Urinary metabolism of exemestane, a new aromatase inhibitor, in rat, dog, monkey and human volunteers. *J Endocrinol Invest* 17: 78
9 Pfister C, Dowsett M, Iveson T et al. (1993) Pharmacokinetics of new aromatase inhibitor CGS 20267. *Eur J Drug Metab Pharmacokinet* 18 (suppl.): 117
10 Spinelli R, Jannuzzo MG, Poggesi I et al. (1999) Pharmacokinetics (PK) of Aromasin (exemestane, EXE) after single and repeated doses in healthy postmenopausal volunteers (HPV). *Eur J Cancer* 35: S295
11 Lønning PE, Bajetta E, Murray R et al. (2000) Activity of exemestane in metastatic breast cancer after failure of nonsteroidal aromatase inhibitors: a phase II trial. *J Clin Oncol* 18: 2234–2244
12 Longcope C (1972) The metabolism of estrone sulfate in normal males. *J Clin Endocrinol Metab* 34: 113–122
13 Murray M, Cantrill E, Farrell GC (1993) Induction of cytochrome P450 2B1 in rat liver by the aromatase inhibitor aminoglutethimide. *J Pharmacol Exp Ther* 265: 477–481
14 Lien EA, Anker G, Lønning PE et al. (1990) Decreased serum concentrations of tamoxifen and its metabolites induced by aminoglutethimide. *Cancer Res* 50: 5851–5857
15 Lundgren S, Lonning PE, Aakvaag A (1990) Influence of aminoglutethimide on the metabolism of medroxyprogesterone acetate and megestrol acetate in postmenopausal patients with advanced breast cancer. *Cancer Chemother Pharmacol* 27: 101–105
16 Van Deijk WA, Blijham GH, Mellink WA, Meulenberg PM (1985) Influence of aminoglutethimide on plasma levels of medroxyprogesterone acetate: its correlation with serum cortisol. *Cancer Treat Rep* 69: 85–90
17 Haynes BP, Dowsett M, Miller WR et al. (2003) The pharmacology of letrozole. *J Steroid Biochem Mol Biol* 87: 35–45
18 Dowsett M, Tobias JS, Howell A et al. (1999) The effect of anastrozole on the pharmacokinetics of tamoxifen in post-menopausal women with early breast cancer. *Br J Cancer* 79: 311–315
19 Dowsett M, Cuzick J, Howell A, Jackson I (2001) Pharmacokinetics of anastrozole and tamoxifen alone, and in combination, during adjuvant endocrine therapy for early breast cancer in postmenopausal women: a sub-protocol of the 'Arimidex and tamoxifen alone or in combination' (ATAC) trial. *Br J Cancer* 85: 317–324
20 Purba HS, Bhatnagar A (1990) A comparison of methods measuring aromatase activity in human placenta and rat ovary. *J Enzyme Inhib* 4: 169–178
21 Brodie A, Lu Q, Liu Y, Long B (1999) Aromatase inhibitors and their antitumor effects in model systems. *Endocr Relat Cancer* 6: 205–210
22 Miller WR, Dixon JM (2000) Antiaromatase agents: preclinical data and neoadjuvant therapy. *Clin Breast Cancer* (suppl.): S9–S14
23 Geisler J, Lønning PE (2005) Aromatase inhibition – translation into a succesful therapeutic approach. *Clin Cancer Res* 11(8): 2809–2821
24 Thompson EA Jr, Siiteri PK (1974) The involvement of human placental microsomal cytochrome P-450 in aromatization. *J Biol Chem* 249: 5373–5378

25 Steele RE, Mellor LB, Sawyer WK et al. (1987) *In vitro* and *in vivo* studies demonstrating potent and selective estrogen inhibition with the nonsteroidal aromatase inhibitor CGS 16949A. *Steroids* 50: 147–161
26 Bossche HV, Willemsens G, Roels I et al. (1990) R 76713 and enantiomers: selective, nonsteroidal inhibitors of the cytochrome P450-dependent oestrogen synthesis. *Biochem Pharmacol* 40: 1707–1718
27 Dukes M, Edwards PN, Large M et al. (1996) The preclinical pharmacology of "Arimidex" (anastrozole; ZD1033) – a potent, selective aromatase inhibitor. *J Steroid Biochem Mol Biol* 58: 439–445
28 Bhatnagar AS, Hausler A, Schieweck K et al. (1990) Highly selective inhibition of estrogen biosynthesis by CGS 20267, a new non-steroidal aromatase inhibitor. *J Steroid Biochem Mol Biol* 37: 1021–1027
29 Brodie AM, Wing LY (1987) *In vitro* and *in vivo* studies with aromatase inhibitor 4-hydroxyandrostenedione. *Steroids* 50: 89–103
30 Di Salle E, Briatico G, Giudici D et al. (1994) Novel aromatase and 5 alpha-reductase inhibitors. *J Steroid Biochem Mol Biol* 49: 289–294
31 Batzl C, Hausler A, Schieweck K et al. (1996) Pharmacology of nonsteroidal aromatase inhibitors. In: Pasqualini J, Katzenellenbogen B (eds): *Hormone-dependent cancer*. Marcel Dekker, New York, 155–168
32 Batzl-Hartmann C, Evans DB, Bhatnagar A (2003) Comparative aromatase enzyme kinetic studies on fadrozole, formestane, letrozole, anastrozole and exemestane. 26$^{th}$ San Antonio Breast Cancer Symposium, San Antonio, TX, USA. Date: December 3–6, 2003. Proceedings published in: Breast Cancer Research and Treatment, Vol. 82, Supplement 1, Abstract 458, page S111, 2003. Kluver Academic Publishers
33 Lønning PE, Jacobs S, Jones A et al. (1991) The influence of CGS 16949A on peripheral aromatisation in breast cancer patients. *Br J Cancer* 63: 789–793
34 Geisler J, Haynes B, Anker G et al. (2002) Influence of letrozole and anastrozole on total body aromatization and plasma estrogen levels in postmenopausal breast cancer patients evaluated in a randomized, cross-over study. *J Clin Oncol* 20: 751–757
35 Kochak GM, Mangat S, Mulagha MT et al. (1990) The pharmacodynamic inhibition of estrogen synthesis by fadrozole, an aromatase inhibitor, and its pharmacokinetic disposition. *J Clin Endocrinol Metab* 71: 1349–1355
36 Santen RJ, Santner S, Davis B et al. (1978) Aminoglutethimide inhibits extraglandular estrogen production in postmenopausal women with breast carcinoma. *J Clin Endocrinol Metab* 47: 1257–1265
37 Lønning PE, Geisler J, Johannessen DC, Ekse D (1997) Plasma estrogen suppression with aromatase inhibitors evaluated by a novel, sensitive assay for estrone sulphate. *J Steroid Biochem Mol Biol* 61: 255–260
38 Jacobs S, Lønning PE, Haynes B et al. (1991) Measurement of aromatisation by a urine technique suitable for the evaluation of aromatase inhibitors *in vivo*. *J Enzyme Inhib* 4: 315–325
39 Dowsett M, Jones A, Johnston SR et al. (1995) *In vivo* measurement of aromatase inhibition by letrozole (CGS 20267) in postmenopausal patients with breast cancer. *Clin Cancer Res* 1: 1511–1515
40 MacNeill FA, Jones AL, Jacobs S et al. (1992) The influence of aminoglutethimide and its analogue rogletimide on peripheral aromatisation in breast cancer. *Br J Cancer* 66: 692–697
41 MacNeill FA, Jacobs S, Dowsett M et al. (1995) The effects of oral 4-hydroxyandrostenedione on peripheral aromatisation in post-menopausal breast cancer patients. *Cancer Chemother Pharmacol* 36: 249–254
42 Jones AL, MacNeill F, Jacobs S et al. (1992) The influence of intramuscular 4-hydroxyandrostenedione on peripheral aromatisation in breast cancer patients. *Eur J Cancer* 28A: 1712–1716
43 Geisler J, King N, Dowsett M et al. (1996) Influence of anastrozole (Arimidex), a selective, nonsteroidal aromatase inhibitor, on *in vivo* aromatisation and plasma oestrogen levels in postmenopausal women with breast cancer. *Br J Cancer* 74: 1286–1291
44 Geisler J, King N, Anker G et al. (1998) *In vivo* inhibition of aromatization by exemestane, a novel irreversible aromatase inhibitor, in postmenopausal breast cancer patients. *Clin Cancer Res* 4: 2089–2093
45 Thürlimann B, Castiglione M, HsuSchmitz SF et al. (1997) Formestane *versus* megestrol acetate

in postmenopausal breast cancer patients after failure of tamoxifen: a phase III prospective randomised cross over trial of second-line hormonal treatment (SAKK 20/90). *Eur J Cancer* 33: 1017–1024

46 Buzdar AU, Smith R, Vogel C et al. (1996) Fadrozole HCL (CGS-16949A) *versus* megestrol acetate treatment of postmenopausal patients with metastatic breast carcinoma. Results of two randomized double blind controlled multiinstitutional trials. *Cancer* 77: 2503–2513

47 Falkson CI, Falkson HC (1996) A randomised study of CGS 16949A (fadrozole) *versus* tamoxifen in previously untreated postmenopausal patients with metastatic breast cancer. *Ann Oncol* 7: 465–469

48 Thürlimann B, Beretta K, Bacchi M et al. (1996) First-line fadrozole HCI (CGS 16949A) *versus* tamoxifen in postmenopausal women with advanced breast cancer – Prospective randomised trial of the Swiss Group for Clinical Cancer Research SAKK 20/88. *Ann Oncol* 7: 471–479

49 Pérez-Carrión R, Candel VA, Calabresi F et al. (1994) Comparison of the selective aromatase inhibitor formestane with tamoxifen as first-line hormonal therapy in postmenopausal women with advanced breast cancer. *Ann Oncol* 5:S19–S24

50 Geisler J, Berntsen H, Lønning PE (2000) A novel HPLC-RIA method for the simultaneous detection of estrone, estradiol and estrone sulphate levels in breast cancer tissue. *J Steroid Biochem Mol Biol* 72: 259–264

51 Geisler J (2003) Breast cancer tissue estrogens and their manipulation with aromatase inhibitors and inactivators. *J Steroid Biochem Mol Biol* 86: 245–253

52 Van Landeghem AA, Poortman J, Nabuurs M, Thijssen JH (1985) Endogenous concentration and subcellular distribution of estrogens in normal and malignant human breast tissue. *Cancer Res* 45: 2900–2906

53 Pasqualini JR, Cortes-Prieto J, Chetrite G et al. (1997) Concentrations of estrone, estradiol and their sulfates, and evaluation of sulfatase and aromatase activities in patients with breast fibroadenoma. *Int J Cancer* 70: 639–643

54 Geisler J, Detre S, Berntsen H et al. (2001) Influence of neoadjuvant anastrozole (Arimidex) on intratumoral estrogen levels and proliferation markers in patients with locally advanced breast cancer. *Clin Cancer Res* 7: 1230–1236

55 Sluijmer AV, Heineman MJ, De Jong FH, Evers JL (1995) Endocrine activity of the postmenopausal ovary: the effects of pituitary down-regulation and oophorectomy. *J Clin Endocrinol Metab* 80: 2163–2167

56 Bolt HM (1979) Metabolism of estrogens-natural and synthetic. *Pharmacol Ther* 4: 155–181

57 Miller WR, Mullen P, Sourdaine P (1997) Regulation of aromatase activity within the breast. *J Steroid Biochem Mol Biol* 61: 193–202

58 de Jong PC, Blankenstein MA, van de Ven J et al. (2001) Importance of local aromatase activity in hormone-dependent breast cancer: a review. *Breast* 10: 91–99

59 Bulun SE, Takayama K, Suzuki T et al. (2004) Organization of the human aromatase p450 (CYP19) gene. *Semin Reprod Med* 22: 5–9

60 Reed MJ, Aherne GW, Ghilchik MW et al. (1991) Concentrations of oestrone and 4-hydroxyandrostenedione in malignant and normal breast tissues. *Int J Cancer* 49: 562–565

61 de Jong PC, van de Ven J, Nortier HW et al. (1997) Inhibition of breast cancer tissue aromatase activity and estrogen concentrations by the third-generation aromatase inhibitor vorozole. *Cancer Res* 57: 2109–2111

62 Miller WR, Telford J, Love CD et al. (1998) Effects of letrozole as primary medical therapy on *in situ* oestrogen synthesis and endogenous oestrogen levels within the breast. *Breast* 7: 273–276

63 Johannessen DC, Engan T, Di Salle E et al. (1997) Endocrine and clinical effects of exemestane (PNU 155971), a novel steroidal aromatase inhibitor, in postmenopausal breast cancer patients: a phase I study. *Clin Cancer Res* 3: 1101–1108

Aromatase Inhibitors
Edited by B.J.A. Furr



# Clinical studies with exemestane

Robert J. Paridaens

*University Hospital Gasthuisberg, Katholieke Universiteit Leuven, Herestraat 49, B-3000 Leuven, Belgium*

## Introduction

### *Background of hormone dependence of breast cancer*

Oestrogen is the major stimulus driving the growth of hormone-dependent breast cancer, and most forms of endocrine therapy are directed towards inhibiting, ablating or interfering with oestrogen activity. In young adult women, the ovary is the main source of oestrogen synthesis, which after a cascade of biochemical steps ultimately occurs by the conversion of androgen precursors (testosterone and androstenedione) into the corresponding oestrogens (oestradiol and oestrone, respectively), specifically mediated through the enzyme, aromatase. Other tissues, like the placenta, muscle, skin and mainly adipose tissue, may also display significant aromatase activity, mediated by tissue-specific isoforms of this enzyme. As ovarian function declines with the onset of the menopause, the relative proportion of oestrogens synthesized in extragonadal sites increases, and eventually non-ovarian oestrogens (mainly oestrone) predominate in the circulation. Enzymatic conversion of androgenic precursors (testosterone and androstenedione), secreted by the adrenals, generates oestradiol and oestrone in peripheral tissues. Aromatase, the enzyme responsible for this conversion, is mainly present in adipose tissue, liver, muscle and brain. Aromatase activity has also been identified in the epithelial and stromal components of the breast. Therefore, local synthesis of oestrogens may contribute to breast cancer growth in postmenopausal women. At the tissue level, complex paracrine and autocrine crosstalk plays an instrumental role in normal breast physiology, but is also crucial for the promotion and development of a cancer. Tumour cells themselves may be able to produce hormones or growth factors, which can promote their own proliferation, or modulate their local environment.

### *Modalities of hormonal therapy*

Beatson's historic publication in 1896 in the *Lancet* [1], reporting breast cancer regression after oophorectomy, was the first scientific proof that an endocrine manipulation may influence the course of the disease. This observation, made long before the identification of the biochemical substrates of hormone dependence (hormones and receptors), led, 50 years later, to the development of other surgical modalities of endocrine ablation like adrenalectomy and hypophysectomy, which were feasible only after hormone-replacement therapy with corticosteroids and thyroid hormone had become available. During the 1960s, successful medical approaches were developed with pharmacological doses of steroids (oestrogens, progestins and androgens) and later antioestrogens, selective oestrogen receptor modulators (SERMs) and aromatase inhibitors, which have now rendered obsolete major endocrine-ablative surgery. Oophorectomy remains in use, but equivalent hormonal suppression of the ovarian endocrine function can be achieved with ovarian irradiation, or with luteinizing hormone-releasing hormone (LHRH) analogues.

## Antioestrogens and SERMs

Tamoxifen, a non-steroidal triphenylethylene, has remained the preferred hormonal treatment for breast cancer over the last four decades. The decline in breast cancer mortality in western countries is considered to be partially due to the use of tamoxifen [2, 3]. After discovery of its antioestrogenic properties in the late 1960s, by showing its ability to bind oestrogen receptor (ER) and to antagonize the effects of oestrogens on cell cultures and in *in vivo* experiments in rodents, the efficacy of tamoxifen has been shown at every stage of the disease. Tamoxifen competes for the binding of oestradiol to the ER, but still allows the dimerization of tamoxifen–receptor complexes, which can interact with the estrogen responsive elements (ERE) at the nuclear level [4]. Tamoxifen retains some oestrogenic agonistic properties on several tissues and organs, like the endometrium and liver, explaining why it can induce endometrium changes (cystic thickening, polyps, growth of fibroids, epithelial hyperplasia and even endometrial carcinoma or sarcoma) and activate the coagulation system with increased propensity for deep-vein thrombosis and stroke [5]. It is also associated with beneficial effects on bone mineral density [6] and blood lipid profile (decrease of the atherogenic fraction of cholesterol), which also represent oestrogenic effects [7]. At the pituitary level, tamoxifen behaves as an antagonist, inducing vasomotor symptoms, sometimes severe and long-lasting. When administered to premenopausal women, tamoxifen can induce multiple ovulations, associated with a marked rise in circulating oestrogens; it can sometimes lead to macro-polycystic changes in the ovaries. The latter complications can be avoided by administering simultaneously an LHRH analogue to block ovarian function. Toremifene, an analogue of tamoxifen, exhibiting the same efficacy and the same safety profile as tamoxifen, over which it has no obvious clinical advantage or disadvantage, is also used. These drugs must be considered as equivalent, and as such also totally cross-resistant [8].

The mixed agonist/antagonist actions of tamoxifen explain several well-described clinical syndromes associated with treatment, like flare-up reactions with hypercalcaemia and bone pain which may occur rapidly, within hours or within a few days after initiation of treatment in patients with bone metastases. Such a flare can be avoided by administering an intravenous bisphosphonate (pamidronate or zoledronate) prior to initiating tamoxifen therapy. Tumour stabilization and, rarely, regression has been described after withdrawal of tamoxifen therapy, indicating that the drug can in fact have an oestrogen-like growth-promoting effect on tumour deposits. The main fear of a clinician prescribing tamoxifen is that the drug may in fact stimulate the tumour by losing its antioestrogenic effect and thus be seen by the tumour cells as purely oestrogenic. Such an oestrogenic switch has been demonstrated in experimental models (cell lines becoming dependent on tamoxifen for their growth), and may be an explanation for the absence of additional beneficial effects by extending adjuvant use of tamoxifen beyond 5 years [9].

Tamoxifen was until recently the standard hormonal therapy for breast cancer patients whose tumours express the ER and/or the progesterone receptor [3]. The development of resistance to tamoxifen in patients with metastatic disease and long-term toxicities, including thromboembolic events and endometrial cancer in patients with early breast cancer, have led to increasing use of alternative hormonal therapies including aromatase inhibitors.

## Steroidal and non-steroidal aromatase inhibitors

Aromatase is the key enzyme that catalyzes oestrogen synthesis by converting androstenedione to oestrone, and testosterone to oestradiol. Inhibition of aromatase reduces circulating oestrogen levels in postmenopausal women, and several trials have shown efficacy of aromatase inhibitors in treating hormone-responsive breast cancer [10]. Inhibition of aromatase is, therefore, an effective strategy for ER-positive, postmenopausal, metastatic breast cancer patients and may be particularly useful when tamoxifen treatment fails.

The first aromatase inhibitors to become clinically available were δ-L-testolactone (Teslac) and aminoglutethimide (Orimeten) [11]. Teslac is a modified androgen, which is believed to compete with androstenedione at the binding site of aromatase. This compound displayed very modest efficacy, and was later replaced by a second-generation steroidal aromatase inhibitor, 4-hydroxyandrostenedione, which unfortunately could only be administered by the intramuscular route [12]. Aminoglutethimide is a non-steroidal aromatase inhibitor without any binding capacity for steroid hormone receptors, which can block aromatization at the level of a cytochrome P450 coenzymatic site. It has demonstrated activity in the metastatic breast cancer setting, eliciting response rates comparable to those achieved by tamoxifen or progestins. Apart from its inhibition of aromatase, it depresses the central nervous system (the drug was initially developed as an anti-convulsant) and can affect other endocrine pathways; it may inhibit glucocorticoid production from the adrenals, and rarely induce hypothyroidism and agranulocytosis. After having been used for about 20 years as second- and third-line endocrine therapy for metastatic disease (after tamoxifen and eventually after progestins), it is now used infrequently in the clinical setting, because it has been replaced by newer aromatase inhibitors that display a much better profile of efficacy and safety.

The latest generation of aromatase inhibitors includes the steroidal compound exemestane as well as the non-steroidal compounds anastrozole and letrozole [12–14]. These newer aromatase inhibitors are superior to aminoglutethimide as well as to megestrol acetate as a second-line modality for treating advanced breast cancer following tamoxifen therapy [15–17]. Like its non-steroidal congeners, the steroidal aromatase inhibitor exemestane has been studied across the spectrum of breast cancer. Exemestane differs from non-steroidal aromatase inhibitors in that it leads to irreversible inhibition of aromatase by covalently binding to the enzyme [13]. Because aromatase inhibitors and aromatase inactivators differ in their mechanisms of action, they are not totally cross-resistant and thus, in clinical practice, represent two distinct classes of drugs.

## Studies with exemestane in metastatic breast cancer

### *Pharmacology and early phase 1/2 studies*

The latest generation of steroidal (exemestane) and non-steroidal (anastrazole, letrozole) aromatase inhibitors acts specifically on peripheral and tumour aromatase and does not suppress adrenal function. By irreversibly (exemestane) or reversibly (anastrazole, letrozole) inhibiting peripheral and tumour aromatase, these drugs are nearly 1000 times more potent than aminoglutethimide, and can reduce levels of circulating oestrogens to undetectable values (with standard assays) in menopausal women, thereby removing very efficiently a growth stimulus for hormone-sensitive tumours [18]. In phase 1, daily doses of exemestane of 0.5–800 mg have been tested [19, 20]. Subjective tolerance was generally excellent, but at doses in excess of 200 mg mild virilization was observed with acne, hoarseness and hirsutism. Therefore, the lower daily dose of 25 mg, at which maximal suppression of circulating oestrogens was obtained, was selected as the recommended dose for further clinical development.

Like tamoxifen, the most frequent side effect reported by postmenopausal women taking aromatase inhibitors remains hot flushes. Many patients also complain of arthralgia and myalgia, but this may be more severe with non-steroidal aromatase inhibitors than with exemestane. Aromatase inhibitors are safe for the uterus: they induce endometrial atrophy and may reverse the changes induced by tamoxifen, as shown by echographic studies [21]. The risk of thromboembolic events during aromatase-inhibitor treatment is substantially lower than for tamoxifen. It is noteworthy that the two classes of aromatase inhibitors – steroidal and non-steroidal – are not totally cross-resistant, and patients failing to respond to one class still have a 25% chance of getting clinical benefit (that is, remission or stable disease for at least 6 months) from the other. Several phase 2 studies have demonstrated the effectiveness of exemestane for advanced breast cancer that has progressed during or after second-line treatment with aminoglutethimide, non-steroidal aromatase inhibitors or megestrol acetate [13, 15, 22, 23]. Conversely, for patients with metastatic disease whose disease progresses on exemestane, recent data indicate that non-steroidal aromatase inhibitors may also be of clinical benefit [24]. As a result, the options available for treating hormonally sensitive breast cancers are expanded; numerous trials have attempted to define the optimal sequence for using the various modalities.

### *Randomized phase 3 studies in second- and first-line treatments*

The efficacy and safety of aromatase inhibitors is already established in all lines of hormonal treatment of postmenopausal patients with metastatic hormone-sensitive tumours. Exemestane proved to be superior to megestrol acetate in prolonging overall survival time, time to tumour progression, and time to treatment failure in a phase 3 study of women with advanced breast cancer who had progressed or relapsed during treatment with tamoxifen [25].

The European Organisation for the Research and Treatment of Cancer (EORTC) has investigated the efficacy and tolerability of exemestane as a first-line therapy for hormone-responsive metastatic breast cancer in postmenopausal women. This was a multicentre, randomized, open-label, phase 2/3 study. Eligible patients were assigned randomly to receive either exemestane at a daily oral dose of 25 mg or tamoxifen at a daily oral dose of 20 mg. Randomization was performed after stratification for institution, previous adjuvant tamoxifen therapy, previous chemotherapy for metastatic disease and dominant site of metastasis (visceral with or without others, bone only, bone and soft tissue, soft tissue only). Patients received the designated treatment until disease progression or unacceptable toxicity; this included patient withdrawal. The initial phase 2 part of this study was designed to assess response rates to exemestane and to determine whether the study should be extended in phase 3 in order to allow a true comparison with tamoxifen [14]. Of patients who received exemestane, 41% achieved an objective response; only 17% responded among those who received tamoxifen. The clinical benefit (proportion of patients achieving a complete response, partial response or disease stabilization) was 57% for exemestane-treated patients and 42% for tamoxifen-treated patients. A low incidence of toxicity was observed. Exemestane was well tolerated, and criteria for trial extension to a phase 3 randomized study were met.

The phase 3 step was designed specifically to compare the efficacy and safety of first-line therapy with exemestane *versus* tamoxifen in terms of progression-free survival. Final results were presented at the ASCO meeting in 2004, and are summarized below. Between October 1996 and December 2002, 382 patients from 81 centres were accrued and randomly assigned to treatment. Approximately 21% of patients in each treatment group had received hormonal therapy previously. The median duration of follow-up was 29 months and was homogeneous across treatments. A total of 319 events (progression or death) were observed in the 371 patients: 161 (85%) in the tamoxifen arm and 158 (87%) in the exemestane arm. The hazard ratio for progression-free survival (PFS) was 0.84 (95% confidence interval (CI), 0.67–1.05) in favour of exemestane. Although the planned log-rank test analysis was not significant ($P = 0.121$), observations of the Kaplan–Meier curves indicated that the hazard ratio did not behave proportionally over time. The median duration of PFS was significantly longer with exemestane than with tamoxifen (10 *versus* 6 months) using the Wilcoxon test ($P = 0.028$). No differences in overall survival were observed between treatment arms and, at 1 year, 82% of tamoxifen- and 86% of exemestane-treated patients had survived. The objective response rate (complete plus partial response) was 46% for the exemestane treatment arm and 31% for the tamoxifen treatment arm. The odds ratio was 1.85 (95% CI, 1.21–2.82; $P = 0.005$; exact $\chi^2$).

The results of the EORTC study are consistent with those observed in other randomized phase 3 studies of aromatase inhibitors and tamoxifen as first-line therapy for metastatic breast cancer. These findings in the metastatic setting support the growing body of evidence that aromatase inhibitors have broad utility throughout the spectrum of breast cancer and may have clinical advantages over tamoxifen in the adjuvant and true preventive setting, as suggested by results comparing anastrozole with tamoxifen [27, 28]. Like exemestane, anastrozole and letrozole have been compared with tamoxifen as first-line treatment [29–32]. All three showed superiority to tamoxifen in hormone-sensitive breast cancer, with significant prolongation of progression-free survival (median PFS is 5–6 months for tamoxifen, and 9–10 months for the aromatase inhibitors) [26, 29–32]. Due to the lack of randomized phase 3 studies comparing steroidal and non-steroidal aromatase inhibitors, it is unknown at this time if any drug is superior to the others.

A companion sub-study of the randomized phase 2 step of the EORTC trial evaluated the impact of exemestane and tamoxifen on the lipid profile of patients by measuring serum triglycerides (TRG), high-density lipoprotein (HDL) cholesterol, total cholesterol (TC), lipoprotein a and apolipoprotein (apo) B and apoA1 at baseline and while on therapy at 8, 24 and 48 weeks [33]. All patients without hypolipidaemic treatment who had baseline and at least one other lipid assessment were included in the analysis; those who received concomitant drugs that could affect lipid profile were included only if those drugs were administered throughout the study treatment. Increases or decreases in lipid parameters within 20% of baseline were considered as non-significant and thus unchanged. Some 72 patients (36 in each arm) were included in the statistical analysis. The majority of patients had abnormal TC and normal TRG, HDL cholesterol, apoA1, apoB and lipoprotein a levels at baseline. Neither exemestane nor tamoxifen had adverse effects on TC, HDL cholesterol, apoA1, apoB or lipoprotein a levels at 8, 24 and 48 weeks of treatment. Exemestane and tamoxifen had opposite effects on TRG levels: exemestane decreased, while tamoxifen increased, TRG levels over time. There were too few patients with normal baseline TC and abnormal TRG, HDL cholesterol, apoA1, apoB and lipoprotein a levels to allow for assessment of exemestane's impact on these sub-sets. The atherogenic risk determined by apoA1/apoB and TC/HDL cholesterol ratios remained unchanged throughout the treatment period in both the exemestane and tamoxifen arms. It was concluded that exemestane had no detrimental effect on cholesterol levels, nor on atherogenic indices, which are well-known risk factors for coronary artery disease. In addition, it had a beneficial effect on TRG levels. These data, coupled with exemestane's excellent efficacy and tolerability, supported further exploration of its potential in the metastatic, adjuvant and chemopreventive settings.

## Adjuvant studies with exemestane

The Intergroup Exemestane Study (IES) trial investigated an original schedule of sequential therapy by randomizing women with hormone-sensitive breast cancer having already received 2–3 years of adjuvant tamoxifen to either pursue the same treatment (2362 patients) or to receive exemestane for 2–3 years (2380 patients), in order to complete a total period of 5 years adjuvant endocrine therapy [34]. This study was prematurely halted by the independent monitoring committee that found, at a planned interim analysis performed with a median follow-up of 30.6 months, that patients given exemestane had better disease-free survival than those given tamoxifen (hazard ratio, 0.68; $P = 0.0005$). The advantage in relapse-free survival in favour of exemestane is estimated to be 4.7% at 3 years after randomization, with a significant reduction in contralateral breast cancers and distant metastatic recurrences. All subgroups of patients regardless of their nodal status (positive or negative) and their receptor status (ER-positive/progesterone receptor-positive or ER-positive/progesterone receptor-negative) had significantly fewer events with exemestane than with tamoxifen. Thromboembolic events were more frequent during tamoxifen treatment, whereas cardiac events, osteoporosis and fractures were more frequent with exemestane. Overall survival was not significantly different in the two groups, with 93 deaths occurring in the exemestane group and 106 in the tamoxifen group.

In the TEAM study, which started later than the IES trial, patients were initially randomized to receive either tamoxifen or exemestane for 5 years postoperatively. The positive IES findings led to a change in the design of TEAM, which is now comparing 5 years of exemestane with 2.5 years of tamoxifen followed by 2.5 years of exemestane. The results of other large-scale, randomized clinical trials investigating the role of non-steroidal aromatase inhibitors in the adjuvant setting have been recently published. All show some advantage of using an aromatase inhibitor either instead of, or after completion of, the 'classical' 5 years adjuvant tamoxifen treatment [27, 35–37], and are reviewed elsewhere in this volume.

## Conclusions and perspectives

For endocrine therapy of metastatic breast cancer, there is still debate over what the optimal sequence of the various hormonal treatments may be, but clearly, in view of their efficacy and safety profile, aromatase inhibitors represent an excellent option for first-line treatment. Tamoxifen may also be safely used as a first-line therapy and one may hope that newer tests will become available to detect tamoxifen resistance. The choice of first-line treatment for metastatic recurrence is, of course, influenced by the kind of adjuvant hormonal therapy prescribed earlier. A short treatment-free interval should preclude the use of the same modality. It may be possible that, just as for the neoadjuvant situation, steroid hormone-responsive tumours co-expressing HER2/neu may be those that should preferentially receive aromatase inhibitors rather than tamoxifen [38], but this remains to be proved in the metastatic situation. After aromatase inhibitors as first-line therapy, the next treatments may then be either tamoxifen, followed by the alternative aromatase inhibitor (steroidal for patients having previously been exposed to non-steroidal, and the converse) or the reverse sequence. The exact place of fulvestrant, a pure antioestrogen devoid of any agonist oestrogenic effect, is still under investigation [39, 40]. Most clinicians would agree that progestins should be used as the last hormonal modality in the sequence, because of their side effects (mainly water retention, weight gain and increased risk of thromboembolism). Well-conducted hormonal therapy, with rational choice of the best modality adapted to the individual patient, contributes to significant prolongation of survival of patients with metastatic disease, with excellent quality of life.

The success of aromatase-inhibitor therapy in postmenopausal women has raised the issue of whether this approach might be successful in premenopausal women. Meta-analysis of first-generation adjuvant trials, run before the era of hormone receptor assays, has clearly shown that postoperative castration had a beneficial effect on disease-free and overall survival, which was maintained after three decades of follow-up [2, 41]. The LHRH agonist goserelin has also been used as a component of adjuvant systemic therapy in early breast cancer. It appears to provide added benefit to cytotoxic chemotherapy, and has the advantage over ovarian ablation of being given for a period of time with return to normal hormonal status when administration is stopped. However, the optimal duration of ovarian suppression in the adjuvant setting is unknown. In more recent randomized studies comparing adjuvant chemotherapy and adjuvant ovarian ablation using either radiation, surgery or an LHRH agonist, with or without tamoxifen, results have failed to show any advantage for chemotherapy [42, 43]. It should also be emphasized that the chemotherapy (intravenous cyclophosphamide, methotrexate and fluorouracil (CMF)) used in these older trials may nowadays be considered as suboptimal according to contemporary criteria that demand, whenever possible, the use of an anthracycline-based chemotherapy in the adjuvant setting. The problem is further complicated by the fact that adjuvant chemotherapy frequently induces ovarian failure, especially in women aged 40 or more.

Unfortunately, inhibition of ovarian aromatase activity in premenopausal women is associated with polycystic ovaries and androgen excess caused by activation of the pituitary-ovarian axis. Thus aromatase-inhibitor therapy as a single modality is contraindicated in premenopausal women. However, consideration is being given to treating premenopausal women who have advanced breast cancer with a combination of ovarian ablation and an aromatase inhibitor, the latter being compared in clinical trials with the combination of ovarian ablation plus tamoxifen in currently running clinical trials. Combining one modality of ovarian ablation with tamoxifen may indeed be considered nowadays as a standard reference treatment for premenopausal women with hormone-responsive breast cancer [44]. Newer-generation adjuvant endocrine studies are investigating the role of combining ovarian ablation with tamoxifen, or with aromatase inhibitors, and address the question of what should be done in young women, including those who continue to menstruate after completion of adjuvant chemotherapy (TEXT, SOFT and PERCHE trials).

The expansion of hormonally based therapeutic options for the treatment of all stages of hormone-sensitive breast cancer is encouraging. Research in progress aimed at fully characterizing the efficacy, safety and tolerability profiles of exemestane and other aromatase inhibitors will help elucidate which agents are most appropriate at each stage of disease as well as the optimal sequence in which they should be given. Numerous other trials are running that aim to define the role of aromatase inhibitors in the adjuvant setting (optimal duration, optimal sequences), or to solve other problems with aromatase inhibitors that, for instance, do not protect the skeleton against postmenopausal bone loss. Attention is now paid to the cardiovascular background of patients, because contrary to tamoxifen, they do not have a preventative effect on myocardial infarction and cerebrovascular thrombosis. Thus prior history of thromboembolic disease may be an argument to prescribe an aromatase inhibitor, while antecedants of coronary or cerebrovascular disease may favour the choice of tamoxifen. The role of tamoxifen and other endocrine therapies in the management of patients with early breast cancer is a rapidly moving field. International guidelines, regularly updated, are available for helping clinicians to make reasonable therapeutic choices in their daily practice [45]. A more exciting alternative is to offer to the patient, whenever possible, the possibility of participating in well-designed clinical trials exploring new drugs or new approaches, or aiming to optimize the so-called standard modalities.

## References


1 Beatson GW (1896) On the treatment of inoperable cases of carcinoma of the mamma: suggestions for a new method of treatment with illustrative cases. *Lancet* ii: 104–107

2 Early Breast Cancer Trialists' Collaborative Group (1988) Effects of adjuvant tamoxifen and cytotoxic therapy on mortality in early breast cancer. An overview of 61 randomized trials among 28,896 women. *N Engl J Med* 319: 1451–1462

3 Jaiyesimi IA, Buzdar AU, Decker DA, Hortobagyi GN (1995) Use of tamoxifen for breast cancer: twenty-eight years later. *J Clin Oncol* 13(2): 513–529

4 Jensen EV, Jordan VC (2003) The oestrogen receptor: a model for molecular medicine. *Clin Cancer Res* 9: 1980–1989

5 Fisher B, Costantino JP, Redmond CK et al. (1994) Endometrial cancer in tamoxifen-treated breast cancer patients: findings from the National Surgicanl Adjuvant Breast and Bowel Project (NSABP) B-14. *J Natl Cancer Inst* 86: 527–537

6 Love RR, Mazess RB, Barden HS et al. (1992) Effects of tamoxifen on bone mineral density in postmenopausal women with breast cancer. *N Engl J Med* 326: 852–856

7 Love RR, Newcomb PA, Wiebe DA et al. (1990) Effect of tamoxifen therapy on lipid and lipoprotein levels in postmenopausal patients with node-negative breast cancer. *J Natl Cancer Inst* 82: 1327–1332

8 Vogel CL, Schemano I, Schoenfelder J et al. (1993) Multicenter phase II efficacy trial of toremifene in tamoxifen-refractory patients with advanced breast cancer. *J Clin Oncol* 11: 345–352

9 Bryant J, Fisher B, Dignam J (2001) Duration of adjuvant tamoxifen therapy. *J Natl Cancer Inst Monogr* 30: 56–61

10 Dixon JM (2004) Exemestane and aromatase inhibitors in the management of advanced breast cancer. *Exp Opin Pharmacother* 5: 307–316

11 Cocconi G (1994) First generation aromatase inhibitors – aminoglutethimide and testololactone. *Breast Cancer Res Treat* 30: 57–80

12 Brodie AM, Garrett WM, Hendrickson JR et al. (1981) Inactivation of aromatase *in vitro* by 4-hydroxy-4-androstene-3,17-dione and 4-acetoxy-4-androstene-3,17-dione and sustained effects *in vivo*. *Steroids* 38 (6): 693–702

13 Jones S, Vogel C, Arkhipov A et al. (1999) Multicenter, phase II trial of exemestane as third-line hormonal therapy of postmenopausal women with metastatic breast cancer. Aromasin Study Group. *J Clin Oncol* 7 (11): 3418–3425

14 Paridaens R, Dirix L, Lohrisch C et al. (2003) Mature results of a randomized phase II multicenter study of exemestane *versus* tamoxifen as first-line hormone therapy for premenopausal women with metastatic breast cancer. *Ann Oncol* 14: 1391–1398

15 Hamilton A, Piccart M (1999) The third-generation non-steroidal aromatase inhibitors: a review of their clinical benefits in the second-line hormonal treatment of advanced breast cancer. *Ann Oncol* 10 (4): 377–384

16 Dombernowsky P, Smith I, Falkson G et al. (1998) Letrozole, a new oral aromatase inhibitor for advanced breast cancer: double-blind randomized trial showing a dose effect and improved efficacy and tolerability compared with megestrol acetate. *J Clin Oncol* 16 (2): 453–461

17 Gershanovich M, Chaudri HA, Campos D et al. (1998) Letrozole, a new aromatase inhibitor: randomized trial comparing 2.5 mg daily, 0.5 mg daily and aminoglutethimide in postmenopausal women with advanced breast cancer. Letrozole International Trial Group. *Ann Oncol* 9: 639–645

18 Geisler J, King N, Dowsett M et al. (1996) Influence of anastrozole (Arimidex), a selective, non-steroidal aromatase inhibitor, on *in vivo* aromatisation and plasma oestrogen levels in postmenopausal women with breast cancer. *Br J Cancer* 74 (8): 1286–1291

19 Lonning PE, Paridaens R, Thurlimann B et al. (1997) Exemestane experience in breast cancer treatment. *J Steroid Biochem Mol Biol* 61 (3–6): 151–155

20 Paridaens R, Thomas J, Wildiers J et al. (1998) Safety, activity and oestrogen inhibition by exemestane in postmenopausal women with advanced breast cancer: a phase I study. *Anticancer Drugs* 9: 673–683

21 Morales L, Timmerman D, Neven P et al. (2005) Third generation aromatase inhibitors may prevent endometrial growth and reverse tamoxifen-induced uterine changes in postmenopausal breast cancer patients. *Ann Oncol* 16: 70–74

22 Thurlimann B, Paridaens R, Serin D et al. (1997) Third-line hormonal treatment with exemestane

in postmenopausal patients with advanced breast cancer progressing on aminoglutethimide: a phase II multicentre multinational study. Exemestane Study Group. *Eur J Cancer* 33 (11): 1767–1773
23 Lonning PE, Bajetta E, Murray R (2000) Activity of exemestane in metastatic breast cancer after failure of nonsteroidal aromatase inhibitors: a phase II trial. *J Clin Oncol* 18: 2234–2244
24 Bertelli G, Garrone O, Merlano M (2002) Sequential use of aromatase inactivators and inhibitors in advanced breast cancer. *ASCO Proceedings* 21: 60a
25 Kaufmann M, Bajetta E, Dirix LY et al. (2000) Exemestane is superior to megestrol acetate after tamoxifen failure in postmenopausal women with advanced breast cancer: results of a phase III randomized double-blind trial. The Exemestane Study Group. *J Clin Oncol* 18 (7): 1399–1411
26 Paridaens R, Therasse P, Dirix L et al. (2004) First-line hormonal treatment for metastatic breast cancer with exemestane or tamoxifen in postmenopausal patients – a randomized phase III trial of the EORTC Breast Group. *ASCO Proceedings* 25: 6
27 Baum M, Buzdar AU, Cuzick J et al. (2002) Anastrozole alone or in combination with tamoxifen *versus* tamoxifen alone for adjuvant treatment of postmenopausal women with early breast cancer: first results of the ATAC randomized trial. *Lancet* 359: 2131–2139; Erratum, *Lancet* (2002) 360: 1520
28 Baum M, Buzdar A, Cuzick J et al. (2003) Anastrozole alone or in combination with tamoxifen *versus* tamoxifen alone for adjuvant treatment of postmenopausal women with early-stage breast cancer: results of the ATAC (Arimidex, Tamoxifen Alone or in Combination) trial efficacy and safety update analyses. *Cancer* 98 (9): 1802–1810
29 Bonneterre J, Thürlimann B, Robertson JFR et al. (2000) Anastrozole *versus* tamoxifen as first-line therapy for advanced breast cancer in 668 postmenopausal women: results of the Tamoxifen or Arimidex Randomized Group Efficacy and Tolerability Study. *J Clin Oncol* 18: 3748–3757
30 Mouridsen H, Gershanovich M, Sun Y et al. (2001) Superior efficacy of letrozole *versus* tamoxifen as first-line therapy for postmenopausal women with advanced breast cancer: results of a phase III study of the International Letrozole Breast Cancer Group. *J Clin Oncol* 19: 2596–2606
31 Mouridsen H, Gershanovich M, Sun Y et al. (2003) Phase III study of letrozole *versus* tamoxifen as first-line therapy of advanced breast cancer in postmenopausal women: analysis of survival and update of efficacy from the International Letrozole Breast Cancer Group. *J Clin Oncol* 21 (11): 2101–2109
32 Nabholz JM, Buzdar A, Pollak M et al. (2000) Anastrozole is superior to tamoxifen as first-line therapy for advanced breast cancer in postmenopausal women: results of a North American multicenter randomized trial. *J Clin Oncol* 18: 3758–3767
33 Atalay G, Dirix L, Biganzoli L et al. (2004) The effect of exemestane on serum lipid profile in postmenopausal women with metastatic breast cancer: a companion study to EORTC Trial 10951, 'Randomized phase II study in first line hormonal treatment for metastatic breast cancer with exemestane or tamoxifen in postmenopausal patients'. *Ann Oncol* 15 (2): 211–217
34 Coombes RC, Hall E, Gibson LJ et al. (2004) A randomized trial of exemestane after two to three years of tamoxifen therapy in postmenopausal women with primary breast cancer. *N Engl J Med* 350: 1081–1092
35 Dowsett M (2003) Analysis of time to recurrence in the ATAC (arimidex, tamoxifen, alone or in combination) trial according to oestrogen receptor and progesterone receptor status. *Breast Cancer Res Treat* 82 (suppl. 1): S7
36 ATAC (Arimidex, Tamoxifen alone or in combination) Trialists' Group (2003) Anastrozole alone or in combination with tamoxifen *versus* tamoxifen alone for adjuvant treatment of postmenopausal women with early-stage breast cancer. *Cancer* 98: 1802–1810
37 Goss PE, Ingle JN, Martino S et al. (2003) A randomized trial of letrozole in postmenopausal women after 5 years of tamoxifen therapy for early-stage breast cancer. *N Engl J Med* 349: 1793–1802
38 Ellis MJ, Coop A, Singh B et al. (2001) Letrozole is more effective neoadjuvant endocrine therapy than tamoxifen for ErbB-1- and ErbB-2-positive, oestrogen receptor-positive primary breast cancer: evidence from a phase III randomized trial. *J Clin Oncol* 18: 3808–3816
39 Osborne CK, Pippen J, Jones SE et al. (2002) Double-blind randomized trial comparing the efficacy and tolerability of Fulvestrant *versus* anastrozole in postmenopausal women with advanced breast cancer progressing on prior endocrine therapy: results of a North American trial. *J Clin Oncol* 20: 3386–3395
40 Howell A, Robertson JF, Quaresma A et al. (2002) Fulvestrant, formerly ICI 182,780, is as effec-

tive as anastrozole in postmenopausal women with advanced breast cancer progressing after prior endocrine treatment. *J Clin Oncol* 20: 3396–3405
41 Early Breast Cancer Trialists' Collaborative Group (1996) Ovarian ablation in early breast cancer: overview of the randomised trials. *Lancet* 348(9036): 1189–1196
42 Castiglione-Gertsch M, O'Neill A, Price KN et al. (2003) Adjuvant chemotherapy followed by goserelin *versus* either modality alone for premenopausal lymph-node negative breast cancer: a randomized trial. *J Natl Cancer Inst* 95: 1833–1841
43 Jonat W, Kaufmann M, Sauerbrei W et al. (2002) Goserelin *versus* cyclophosphamide, methotrexate, and fluorouracil as adjuvant therapy in premenopausal patients with node-positive breast cancer: the Zoladex Early Breast Cancer Research Association Study. *J Clin Oncol* 20: 4628–4637
44 Klijn JG, Blamey RW, Boccardo F et al. (2001) Combined tamoxifen and luteinizing hormone-releasing hormone (LHRH) agonist *versus* LHRH agonist alone in premenopausal advanced breast cancer: a meta-analysis of four randomized trials. *J Clin Oncol* 19: 343–350
45 Goldhirsch A, Wood WC, Gelber RD et al. (2003) Meeting highlights: updated international expert consensus on the primary therapy of early breast cancer. *J Clin Oncol* 21: 3357–3365

Aromatase Inhibitors
Edited by B.J.A. Furr



# Clinical studies with letrozole

J. Michael Dixon

*Edinburgh Breast Unit, Western General Hospital, Crewe Road, Edinburgh EH4 2XU, UK*

## Introduction

Breast cancer is the most common malignancy in women and a leading cause of cancer death [1]. In 1998, approximately 315,000 women died of breast cancer: nearly two-thirds of these women were postmenopausal [2]. Current treatment options for breast cancer depend on disease characteristics (e.g. stage, sites of any metastases, hormone receptor status), patient characteristics (e.g. age, menopausal status) and patient preferences. Early breast cancer is usually treated with a combination of local (surgery/radiation) and systemic (cytotoxic/endocrine) therapies. Women with inoperable or large operable tumours may be given preoperative or neoadjuvant therapy to shrink the tumours before surgery. Following tumour removal, patients generally receive adjuvant chemotherapy and/or endocrine therapy to reduce the risk of recurrence. Tamoxifen remains the most widely used adjuvant endocrine treatment in women with hormone-responsive tumours. However, following 5 years of adjuvant tamoxifen treatment, patients remain at substantial risk of recurrence [3]. In fact, most breast cancer recurrences and deaths occur more than 5 years after diagnosis and primary adjuvant treatment [3].

Due to their long-term efficacy and good tolerability, endocrine agents are the mainstay for treatment of hormone receptor-positive metastatic, or advanced, breast cancer. In this setting, treatment is aimed at relieving symptoms, delaying progression and improving survival.

The clinical rationale behind endocrine therapies is to deprive the tumour of oestrogen, which is the major established mitogen for human breast cancer *in vivo* [4]. Among women with oestrogen receptor-positive (ER+) or progesterone receptor-positive (PgR+) tumours, 50–60% will respond to initial endocrine therapy [5].

Letrozole (Femara®; Novartis Oncology) is a selective, competitive, nonsteroidal aromatase inhibitor. In postmenopausal women, the conversion of adrenal androgen to oestrogen by aromatase in peripheral tissue is the major source of circulating oestrogen [6–8]. Aromatase activity is present in many tissues throughout the body including the ovaries, adipose tissue, liver, brain, breast and muscle [8]. The mode of action of the aromatase inhibitors differs from that of the antioestrogen tamoxifen in that, whereas antioestrogens compete with the natural ligand for binding to the ER, aromatase inhibitors prevent oestrogen biosynthesis [9, 10]. Letrozole is a highly specific aromatase inhibitor and does not cause the range of side effects associated with inhibition of adrenal corticosteroid synthesis seen with less specific inhibitors such as aminoglutethimide.

In all trials published to date, letrozole has proven superior in one or more aspects to the previous standard of care. It is the only agent to be tested and to confer a benefit in the extended adjuvant setting post-tamoxifen, and the first aromatase inhibitor to demonstrate an overall survival benefit in an adjuvant trial, although this benefit was only seen in women with node-positive disease [11, 12]. Letrozole compared favourably with the first-generation aromatase inhibitor, aminoglutethimide [13], and induced a higher objective response rate (complete plus partial responses, ORR) than anastrozole ($P = 0.013$) in a direct comparison in the second-line setting in advanced breast cancer (Tab. 1) [14]. While this difference was seen in the intent-to-treat population and in defined subgroups with receptor status unknown, soft-tissue or visceral-dominant disease, there was no difference in response rate in women with hormone receptor-positive disease [14].

Letrozole has been used for primary systemic (neoadjuvant) treatment of locally advanced, hormone receptor-rich breast cancer characterised by large (≥T2) or large operable tumours. In a multicentre neoadjuvant trial, letrozole proved superior to tamoxifen in ORR determined by clinical assessment, mammography and ultrasound [15]. Compared with tamoxifen, letrozole enabled more patients to undergo breast-conserving surgery at the end of the treatment period.

Letrozole is currently being investigated as early adjuvant therapy in the Breast International Group 1-98 (BIG 1-98) trial. In this study, letrozole for 5 years is being compared directly with tamoxifen for 5 years. In addition, two further arms are investigating the efficacy of letrozole-tamoxifen sequences during the 5-year early adjuvant period: letrozole for 2 years followed by tamoxifen for 3 years and tamoxifen for 2 years followed by letrozole for 3 years (Fig. 1). Early results suggest that starting adjuvant therapy with letrozole gives a significant improvement in disease-free survival (DFS) and time to recurrence compared with starting with tamoxifen [16].

Table 1. Efficacy outcomes in a comparative trial of letrozole *versus* anastrozole [14]

| | Letrozole | Anastrozole | *P* value |
|---|---|---|---|
| Objective tumour response* | 68 (19%) | 44 (12%) | 0.013 |
| Median TTP | 5.7 weeks | 5.7 weeks | 0.920 |
| Median overall survival | 22 months | 20 months | 0.624 |

*Patients with confirmed complete responses (CR) and partial responses (PR).TTP, time to progession. Analysis based on Cochran–Mantel–Haenszel methodology.

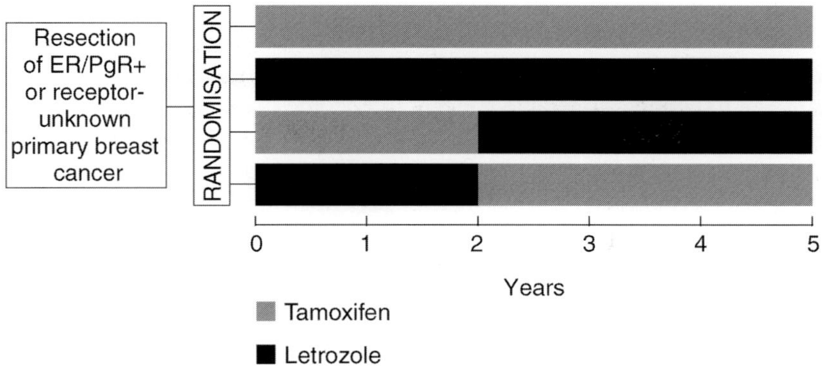


Figure 1. Design of study BIG 1-98 comparing letrozole and tamoxifen in the early adjuvant setting [16].

The extended adjuvant MA.17 trial established that treatment with letrozole following standard adjuvant tamoxifen therapy in postmenopausal women with early breast cancer significantly reduced the risk of recurrence, irrespective of nodal status, and conferred a statistically significant survival advantage in women with node-positive tumours [11, 12]. The side-effect profiles of letrozole and placebo were similar in this study, with no significant differences in discontinuation of therapy, or incidence of cardiovascular events or fractures, although there was a small but statistically significant increase in new-onset, patient-reported osteoporosis [12]. Letrozole is now licensed in this novel setting, offering effective adjuvant therapy for longer than the 5-year limit imposed by the risk:benefit characteristics of tamoxifen.

In advanced breast cancer, letrozole has been used in the first- and second-line settings. In the first-line treatment of postmenopausal women with hormone receptor-positive or -unknown locally advanced or metastatic breast cancer, letrozole proved superior to tamoxifen with regard to time to progression (TTP), ORR and clinical benefit rate, in the largest first-line trial conducted to date [17, 18]. Letrozole was also superior to tamoxifen in terms of 1-year and 2-year survival rates.

In the second-line setting, letrozole has proved superior in at least one endpoint to megestrol acetate [19], aminoglutethimide [13] and anastrozole [14]. Compared with megestrol acetate, letrozole achieved a greater ORR and significantly longer median duration of response [19]. Compared with aminoglutethimide, letrozole was associated with improved TTP and overall survival [13]. In a head-to-head comparison with anastrozole, letrozole demonstrated a significantly higher ORR than anastrozole, although there were no differences in TTP and overall survival (Tab. 1) [14]. The extent of aromatase inhibition and suppression of oestrogen synthesis in patients with advanced breast cancer has also been shown to be greater with letrozole compared with anastrozole [20].

## Primary systemic therapy in early breast cancer

Preoperative, or neoadjuvant, chemotherapy has been used to produce tumour shrinkage to enable inoperable cancers to become operable and patients with large cancers that would require mastectomy to become eligible for breast-conserving surgery. However, in postmenopausal women who are either unfit for, or reject chemotherapy, and in those with ER-rich tumours, endocrine therapy has been used. Early use of tamoxifen gave many women the opportunity to become candidates for breast-conserving surgery instead of mastectomy. The role of letrozole in this setting was initially investigated in a phase II study in 24 patients, which found that preoperative letrozole reduced tumour volume (based on clinical measurements) by an average of 81%, rendering all 24 patients eligible for breast-conserving surgery [21].

As a consequence of these promising results, a double-blind, multicentre, phase IIb/III P024 study was initiated in 337 postmenopausal patients with breast cancer. Patients were randomly assigned to letrozole 2.5 mg/day or tamoxifen 20 mg/day for 4 months prior to surgery [15]. Patients had primary, untreated ER+ and/or PgR+ breast cancer, with clinical stage T2–T4 tumours, nodal stage N0, N1, or N2, without metastases (M0). Patients were not eligible for breast-conserving surgery at the time of presentation. Of the 337 patients enrolled, 154 patients in the letrozole arm and 170 in the tamoxifen arm were included in the intent-to-treat efficacy analysis. Treatment arms were well balanced for baseline characteristics.

The primary endpoint of the P024 study was the percentage of patients in each treatment arm with objective responses (complete or partial response) determined by clinical palpation of the breast cancer. Secondary endpoints were overall ORR determined by mammography and ultrasound at 4 months, and the percentage of patients in each treatment arm who became eligible for breast-conserving surgery. World Health Organization response criteria based on bidimensional measurements of area were applied. All efficacy endpoints showed statistical superiority in favour of letrozole [15].

### *Clinical results*

Significantly more letrozole-treated patients had an objective clinical response compared with tamoxifen-treated patients (55% *versus* 36%; $P < 0.001$). The superiority of letrozole was observed irrespective of baseline tumour size (T2 *versus* >T2) [15].

### *Ultrasound and mammographic response rates*

Letrozole was significantly more effective than tamoxifen irrespective of the assessment method, although response rates assessed by ultrasound and mammography were lower than those assessed by clinical examination. The ORRs for letrozole and tamoxifen, respectively, were 35% *versus* 25% ($P = 0.042$) when assessed by ultrasound, and 34% *versus* 16% ($P < 0.001$) when assessed by mammography (Fig. 2) [15, 22]. Letrozole was also superior to tamoxifen in the subgroup of patients with tumours >T2. When assessed by ultrasound, 38% of patients with tumours >T2 treated with letrozole had an objective response compared with 17% of tamoxifen-treated patients. The difference for mammographic response was even greater in these larger tumours, with letrozole- and tamoxifen-treated patients showing responses of 42% and 18%, respectively [22].

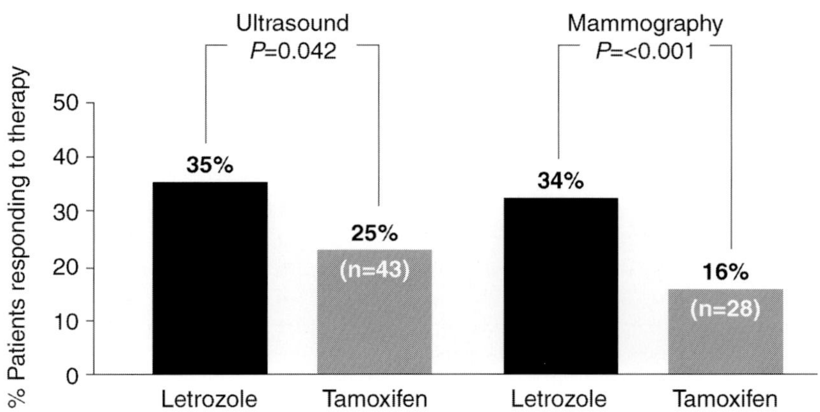


Figure 2. Clinical response by ultrasound and mammography. Independent of measuring technique, letrozole proved superior to tamoxifen [15, 22].

## *Rate of breast-conserving surgery*

The higher response rates assessed by clinical examination were reflected by significantly more letrozole-treated patients than tamoxifen-treated patients being suitable for, and undergoing, breast-conserving surgery (45% *versus* 35%; $P = 0.022$) [15]. Even in patients with locally advanced breast cancer, significantly more patients from the letrozole arm than from the tamoxifen arm were eligible for breast-conserving surgery [22]. At the end of therapy, 135 (88%) patients in the letrozole arm underwent some type of surgery, compared with 139 (82%) patients in the tamoxifen arm.

## *Clinical response analysis*

An exploratory analysis investigating the association between baseline variables (treatment allocation, tumour size, nodal involvement, age) and response showed that the only factor influencing clinical response was the type of therapy used. The odds ratio for treatment was 2.23 (95% confidence interval (CI), 1.43 to 3.50; $P = 0.0005$), indicating that the odds of achieving a response were more than twice as high with letrozole than with tamoxifen [15].

In the exploratory analysis for breast-conserving surgery, baseline tumour size was the most important predictive variable. The odds of undergoing breast-conserving surgery were 4.5 times higher for patients with T2 tumours than for patients with T3 or T4 tumours. Apart from tumour size, the only other factor that influenced the rate of breast-conserving surgery was treatment. The odds of undergoing breast-conserving surgery were increased by more than 70% with letrozole compared with tamoxifen (Tab. 2) [15, 22].

Table 2. Exploratory analysis of breast-conserving surgery. Tumour size and choice of treatment are significant predictors [15, 22]

| Variable | Odds ratio | 95% CI | *P* value |
|---|---|---|---|
| Treatment (letrozole *versus* tamoxifen) | 1.71 | 1.06–2.78 | 0.03 |
| Baseline tumour size (T2 *versus* >T2) | 4.56 | 2.75–7.55 | 0.0001 |
| Nodal involvement (yes *versus* no) | 1.16 | 0.71–1.90 | 0.56 |
| Age (<70 *versus* ≥70 years) | 0.86 | 0.53–1.41 | 0.56 |

An odds ratio >1 favours the underlined variable.

### *Response related to tumour oestrogen receptor expression*

The P024 neoadjuvant study provided an opportunity to investigate the relationship between ER expression levels and response rates in more detail [23]. The histopathological Allred score adds the scores based on intensity of ER expression (range 0–3) and percentage of positive cells (range 0/1–5) [24].

Comparing letrozole and tamoxifen in the neoadjuvant setting, letrozole response rates were numerically superior to tamoxifen response rates in all Allred categories from 3 to 8. This observation indicates that letrozole is more effective than tamoxifen regardless of the level of expression of ER. However, in patients whose tumours had low ER expression (Allred scores 3–5), responses were only achieved with letrozole (Fig. 3) [23].

The response to letrozole in tumours with low ER expression levels suggests that some women who have not previously benefited from standard endocrine therapy due to low ER expression could potentially benefit from treatment with letrozole. This observation could explain some of the differences seen in trial results of different aromatase inhibitors and may have implications for the future choice of adjuvant endocrine agents in these women.

In summary, letrozole is effective in postmenopausal women as neoadjuvant therapy for ER+ and/or PgR+ primary breast cancer and is significantly better

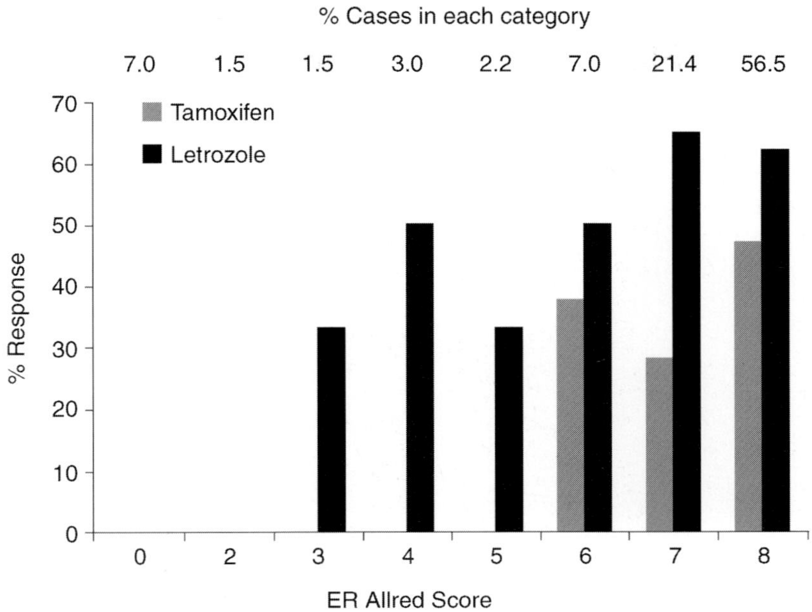


Figure 3. Clinical response rate *versus* ER Allred score for letrozole and tamoxifen. The *P* value for a linear logistic model was 0.0013 for letrozole and 0.0061 for tamoxifen according to the Wald test. In this analysis, ER–/PgR+ cases were excluded. Reproduced with permission [23].

than tamoxifen in reducing tumour size and achieving operability. Furthermore, letrozole is particularly effective compared with tamoxifen (with respect to response rates) in low ER-expressing tumours. The greater efficacy of letrozole compared with tamoxifen in endocrine treatment-naïve tumours suggests that letrozole will also prove more effective than tamoxifen in the adjuvant setting post-surgery.

### *Duration of neoadjuvant letrozole therapy*

A study of 142 postmenopausal women with large operable or locally advanced ER-rich (Allred score ≥5) breast cancer assessed response to letrozole 2.5 mg/day during months 0–3, 3–6 and 6–12 [25]. The median reduction in tumour volume as measured by ultrasound was 46% during months 0–3, an additional 46% during months 3–6, and a further 39.5% during months 6–12 (Fig. 4). This study showed that 3–4 months treatment with letrozole, which is used in most studies of neoadjuvant letrozole, may not be the optimum duration, and that longer durations produced greater tumour shrinkage. Treatment periods of 6 months or longer should increase the numbers of patients with a complete clinical response and the numbers whose disease is downstaged.

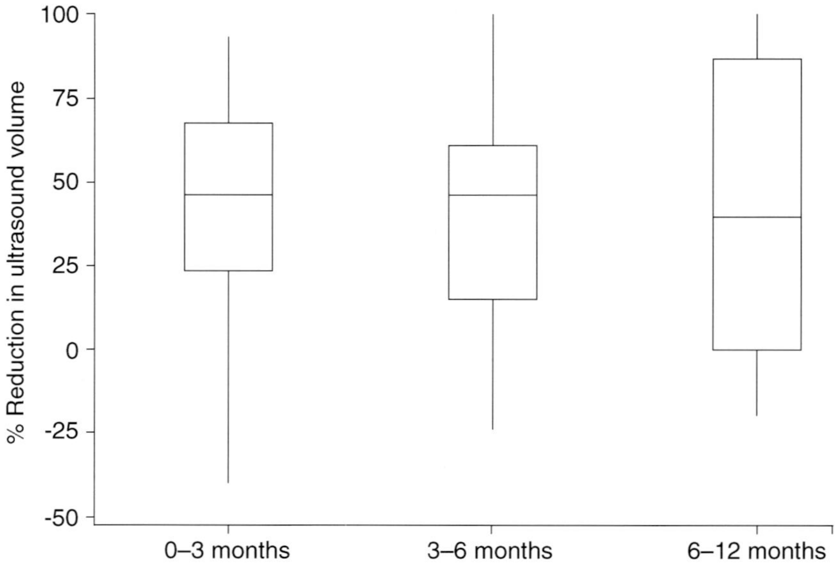


Figure 4. Reduction in ultrasound volume of tumours from postmenopausal women with large operable or locally advanced breast cancer during three time periods. Plots are median and interquartile ranges with outliers [25].

## Clinical trials in progress in the adjuvant setting

### *BIG 1-98*

The BIG 1-98 is a randomised, double-blind, controlled trial that had enrolled more than 8000 postmenopausal patients by closure of recruitment in May 2003 and will provide guidance on the optimal use of letrozole specifically, and aromatase inhibitors in general, in the early adjuvant setting [16].

BIG 1-98 is the only adjuvant trial to compare aromatase inhibitor monotherapy with tamoxifen, as well as comparing both agents used sequentially: tamoxifen followed by letrozole and letrozole followed by tamoxifen. It is also the only aromatase inhibitor trial to prospectively randomise patients to sequential adjuvant treatment immediately post-surgery, rather than after a 2–3-year recurrence-free interval on tamoxifen.

Patients have been randomised into four treatment arms following surgery, as follows:

- letrozole 2.5 mg once daily for 5 years (n = 2400)
- tamoxifen 20 mg once daily for 5 years (n = 2400)
- tamoxifen 20 mg once daily for 2 years crossed over to letrozole 2.5 mg once daily for 3 years (n = 1500)

- letrozole 2.5 mg once daily for 2 years crossed over to tamoxifen 20 mg once daily for 3 years (n = 1500).

Only patients with ER+ and/or PgR+ tumours were enrolled in the trial. The prospectively defined clinical endpoints include DFS (primary endpoint), distant and local-regional DFS, overall survival, and safety. The trial is designed to show superiority over tamoxifen (Fig. 1). The primary core analysis comparing first-line letrozole and tamoxifen included patients from all treatment arms: in the sequential arms, events that occurred more than 30 days after crossover were excluded from the analysis. The median follow-up was 25.8 months, with over 1200 patients being followed for more than 5 years.

Letrozole was shown to significantly increase DFS (hazard ratio 0.81; $P = 0.003$) compared with tamoxifen, and to reduce the risk of relapse at distant sites by 27%; $P = 0.016$), which is a well-recognised predictor of breast cancer death. Time to recurrence (hazard ratio 0.72; $P = 0.0002$) and time to distant metastasis (hazard ratio 0.73; $P = 0.0012$) were also significantly greater in patients receiving letrozole than those receiving tamoxifen. Significantly fewer first-failure events occurred in patients receiving letrozole at local (P = 0.047) and distant (P = 0.006) sites, and the cumulative incidence of breast cancer deaths demonstrated a 3.4% difference in favour of letrozole at 5 years from randomization (P = 0.0002). Letrozole appeared of particular benefit compared with tamoxifen in patients with node-positive disease (hazard ratio 0.71) and patients who had previously received chemotherapy (hazard ratio 0.70) [16].

Current follow-up has not revealed a statistically significant difference in overall survival with letrozole compared with tamoxifen (hazard ratio 0.86; $P = 0.16$) [16]. However, as the benefit with letrozole is likely to be cumulative during treatment, longer follow-up is required to assess any significant effect on mortality.

Data from the crossover arms of the BIG 1-98 study will provide important information on the use of letrozole in sequential treatment strategies with tamoxifen in the adjuvant setting.

*Side-effect profile*

The side-effects that have been reported in patients receiving first-line letrozole therapy for early breast cancer are consistent with oestrogen deficiency resulting from administration of this class of drugs. However, the follow-up in BIG 1-98 is still relatively short, and further data on long-term toxicities will become available in subsequent years. The tolerability of letrozole was shown to be comparable to that of tamoxifen despite differences in toxicity profiles. Slightly more patients on tamoxifen than on letrozole reported at least one serious adverse event (587 *versus* 643, respectively). Patients receiving tamoxifen had significantly more grade 3–5 thromboembolic episodes (odds ratio 0.38; $P < 0.0001$) and a higher incidence of gynaecological events. A trend for fewer cases of invasive endometrial cancer was seen in patients receiving letrozole (odds ratio 0.4; $P = 0.087$). In contrast, letrozole therapy was associated with a higher incidence of fractures (odds ratio 1.42; $P = 0.0006$), and musculoskeletal events, including arthralgia and myalgia [16]. Hypercholesterolaemia was significantly more common in patients receiving letrozole, but this observation was based on non-fasting measurements, and >80% of all reported incidents were grade 1 [16]. Further analysis of these data is pending.

Overall, fewer deaths occurred on-study in patients receiving letrozole than tamoxifen (166 *versus* 192) [16], however letrozole therapy was associated with slightly more deaths without a prior cancer event, but this difference was not statistically significant (55 [1.3%] *versus* 38 [0.9%]; $P = 0.08$). The differences were in cerebrovascular (7 *versus* 1) and cardiac (26 *versus* 13) deaths.

Tamoxifen protects against bone loss, and has cardioprotective properties and favourable effects on serum lipid profiles, so clinical trials comparing an aromatase inhibitor with tamoxifen may not reflect aromatase inhibitor toxicity profiles so much as the difference between aromatase inhibitor toxicity and the beneficial effects of tamoxifen. Consistent with this suggestion, no detrimental effect on cardiovascular disease was seen in the placebo-controlled randomised trial comparing 5 years of letrozole after 5 years of tamoxifen adjuvant therapy with no further therapy (see below) [11]. Recently reported results from the MA.17 lipid substudy (MA.17L) have also not shown any detrimental effect of letrozole compared with placebo on lipid profiles [26].The effects of letrozole on the cardiovascular system have yet to be fully determined, and further follow-up is required to determine the significance of these observations from adjuvant trials.

The overall incidence of grade 3–5 cardiovascular adverse events was similar in letrozole- and tamoxifen-treated patients. Fewer patients receiving letrozole experienced grade 3–5 venous thromboembolic events (0.8% *versus* 2.1%, $P < 0.0001$), but more patients experienced grade 3–5 cardiac events (2.1% *versus* 1.1%); however, the overall numbers of cardiovascular adverse events were small.

### *Z-FAST/ZO-FAST*

All trials assessing aromatase inhibitor use in the adjuvant setting published to date have demonstrated a detrimental effect of these agents on bone mineral density [11, 16, 27, 28]. This effect is almost certainly related to the near-complete oestrogen depletion achieved by aromatase inhibitors, and occurs irrespective of the steroidal/non-steroidal nature of the drug.

Postmenopausal bone loss and its potential consequences can be treated, if not prevented. International guidelines have already addressed this issue [29]. One class of agents that can help to manage cancer treatment-induced bone loss are the bisphosphonates. Within the Z/ZO-FAST trial programmes, the potent bisphosphonate zoledronic acid is used either immediately, or as a delayed therapeutic intervention in the presence of demonstrable bone loss, in patients with early breast cancer receiving adjuvant letrozole therapy. The aim of these trials is to assess the occurrence of bone loss during adjuvant aromatase inhibitor therapy and define the best therapeutic approach to limit this effect. The ZO-FAST and Z-FAST trials have recruited more than 1000, and more than 600, postmenopausal women, respectively. All are patients with stage I–IIIa, ER+ and/or PgR+ breast cancer starting therapy with letrozole, 2.5 mg/day, for 5 years: ZO-FAST closed recruitment at the end of 2004. In both studies, patients were randomised to receive either immediate or delayed zoledronic acid, 4 mg by i.v. infusion every 6 months. Delayed treatment with zoledronic acid is started when the post-baseline T-score decreases by more than 2 standard deviations, or clinical fracture occurs, or if there is evidence of asymptomatic fracture at 36 months. The data from these two trials will be combined. The primary endpoint of both the Z-FAST and ZO-FAST trials is the percentage change in lumbar spine bone mineral density at 12 months.

Preliminary 6-month results from the Z-FAST trial revealed a 1.55% gain in bone mass at the lumbar spine in women assigned to receive upfront zoledronic acid and a 1.78% reduction in bone mass in those assigned to receive delayed zoledronic acid, equivalent to a 3.3% improvement in bone mass for upfront treatment compared with delayed treatment [30]. Thus, upfront zoledronic acid may be able to prevent bone loss in women receiving adjuvant aromatase inhibitor therapy. Further results from these trials will answer important questions on the use of bisphosphonates with aromatase inhibitors and will provide information on the benefits of bisphosphonates in the adjuvant setting.

## Extended adjuvant therapy in early breast cancer

Although tamoxifen is currently being challenged by modern aromatase inhibitors, it remains the standard adjuvant endocrine therapy for women with hormone-responsive early breast cancer following local management of the primary tumour. However, while 5 years of tamoxifen treatment has been shown to improve significantly disease-free and overall survival, the beneficial effects of this agent are limited [3].

Early breast cancer can be considered a chronic disease; patients with all stages of primary breast cancer are at a substantial and continuing risk of relapse following completion of 5 years of adjuvant therapy with tamoxifen, even in the absence of lymph node involvement [31, 32]. In fact, more than 50% of breast cancer relapses and deaths occur after the completion of adjuvant therapy (Fig. 5) [3]. Extending tamoxifen beyond 5 years to address this continuing risk of late recurrence has not proven beneficial. In fact, this approach resulted in an increasing risk of endometrial cancer and other serious side effects and had a detrimental effect on DFS [33].

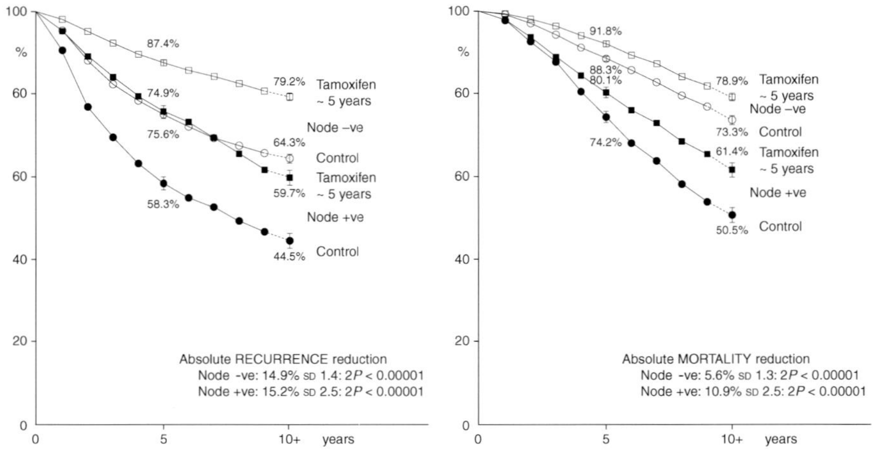


Figure 5. Absolute risk reductions in breast cancer recurrence and mortality during the first 10 years following diagnosis in control patients and patients receiving 5 years of tamoxifen therapy. Women with ER-poor disease were excluded. The values at 5 years and 10 years are given beside each pair of lines and differences in 10-year outcomes are given below the lines. Reproduced with permission [3].

## *Extended adjuvant trial of letrozole* versus *placebo after standard tamoxifen (MA.17 trial)*

A large, randomised, double-blind, placebo-controlled phase III trial compared letrozole and placebo as extended adjuvant therapy in postmenopausal women with hormone-sensitive early breast cancer following standard adjuvant tamoxifen therapy. The aim of the trial was to determine whether, following approximately 5 years of adjuvant tamoxifen therapy, extending adjuvant treatment with letrozole for another 5 years would provide benefits in outcome compared with no further treatment [11].

Postmenopausal women (n = 5157) with ER+ and/or PgR+ or receptor-unknown early breast cancer were recruited to this study (Fig. 6) [11]. Prospective stratification of patients was performed according to receptor status, nodal status and prior chemotherapy. Most patients had hormone receptor-positive disease (98%), approximately half were node-positive and half node-negative, and 46% had received prior adjuvant chemotherapy [11]. The two treatment arms were well balanced for all demographic parameters.

Extended adjuvant treatment with letrozole 2.5 mg daily was initiated within 3 months following completion of 4.5–6 years of adjuvant tamoxifen, in the absence of any disease recurrence. The primary endpoint of MA.17 was DFS, defined as the time to recurrence of the original cancer – either locally, in regional nodes, or as distant metastases – or to the occurrence of a new contralateral breast primary cancer. Secondary endpoints included overall survival, safety, and quality of life. MA.17 companion studies are evaluating treatment effects on bone mineral density (n = 226) and lipid levels (n = 347) [11].

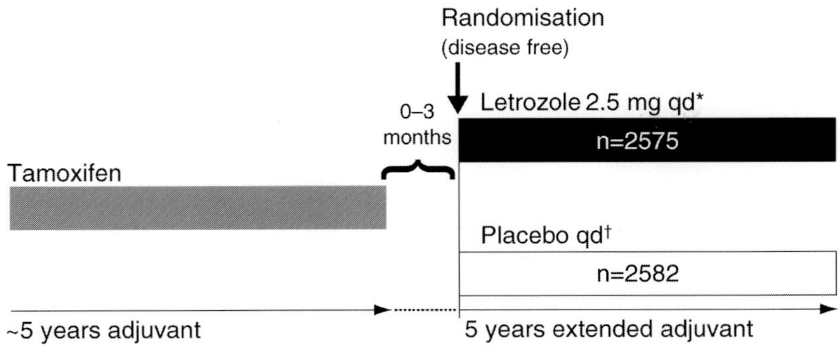


*n=2575 (efficacy); 2154 (safety) in the letrozole arm
†n=2582 (efficacy); 2145 (safety) in the placebo arm

Figure 6. Design of trial MA.17: extended adjuvant letrozole *versus* placebo [11].

According to pre-defined stopping criteria, the trial was unblinded at the first interim analysis due to a significant difference in total events that was shown to favour the letrozole arm [11]. Final analysis of efficacy data was at a median follow-up of 2.5 years, when a total of 247 events and 113 deaths had been observed [12]. For the primary endpoint of DFS, progressive improvement was seen with letrozole *versus* placebo with each year of treatment, and final estimated 4-year DFS was significantly higher for letrozole (4.8% absolute improvement; hazard ratio 0.58; $P = 0.00004$) (Fig. 7). Letrozole reduced the overall risk of recurrence by 42%, and the risk of developing distant metastases was reduced by 40% [11, 12].

Letrozole significantly improved DFS irrespective of prior chemotherapy or nodal status. In node-positive patients, letrozole not only reduced the incidence of distant metastases, but also improved overall survival significantly, reducing mortality by 39% ($P = 0.04$). This is the only significant improvement in overall survival seen in any adjuvant trial of aromatase inhibitors to date. At 30 months of median follow-up, a significant overall survival benefit was not apparent in node-negative patients, but the reduction in local recurrences, distant recurrences, and new primaries in node-negative patients was similar to that seen in patients with nodal involvement [11, 12].

*Side-effect profile*

Letrozole had a similar side-effect profile to placebo in the extended adjuvant setting (Tab. 3) [11, 12]; discontinuation of therapy was not significantly different between the letrozole and placebo groups [11]. The incidence of fractures was not significantly different between letrozole and placebo (5.3% *versus* 4.6%, respectively), but there was a small but significant increase in newly-diagnosed, patient-reported osteoporosis (8% letrozole *versus* 6% placebo,

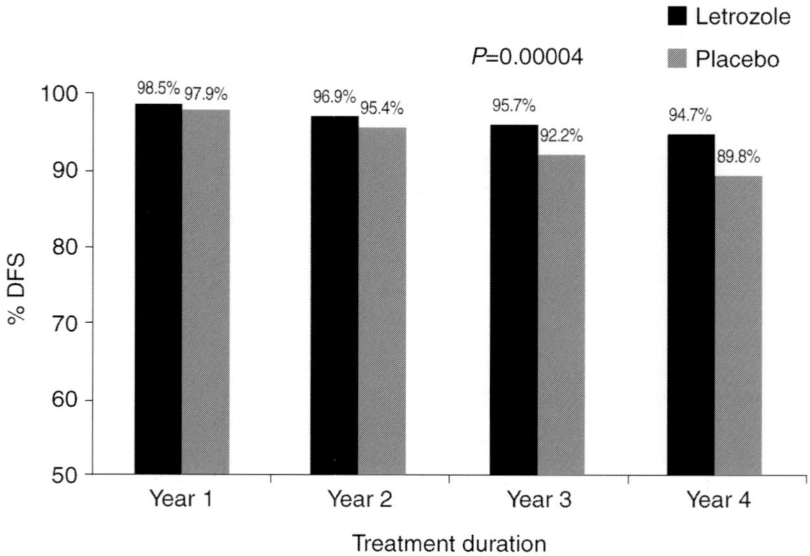


Figure 7. Progressive improvement in DFS with letrozole *versus* placebo with extended adjuvant treatment [11].

$P = 0.003$) [12]. However, in the bone sub-study (MA.17B) of this trial, the incidence of newly diagnosed osteoporosis based on T-score measurement was lower than patient-reported osteoporosis in both treatment arms (3.3% letrozole *versus* 0% placebo): this difference between treatment groups did not reach statistical significance [34].

Table 3. Adverse events of any grade for letrozole *versus* placebo [11, 12]

| Adverse events* | % of patients | | |
|---|---|---|---|
| | Letrozole (n = 2563) | Placebo (n = 2573) | *P* value |
| Hot flushes | 58 | 54 | 0.003 |
| Arthralgia/arthritis | 25 | 21 | <0.0001 |
| Myalgia | 15 | 12 | 0.04 |
| Vaginal bleeding | 6 | 8 | 0.005 |
| Hypercholesterolaemia | 16 | 16 | 0.79 |
| Cardiovascular events | 6 | 6 | 0.76 |
| Osteoporosis (patient-reported new diagnoses) | 8 | 6 | 0.003 |
| Clinical fractures | 5 | 5 | 0.25 |

* 90% of all adverse events were grade 1 or 2.

Letrozole was not associated with any increase in the incidence of cardiovascular events (4.1% *versus* 3.6%; $P = 0.4$) or hypercholesterolaemia (11.9% *versus* 11.5%; $P = 0.67$) compared with placebo [11]. Although data from the BIG 1-98 study indicated that letrozole may be associated with hypercholesterolaemia, data from the extended adjuvant setting do not support this suggestion. In the MA.17L lipid sub-study, no differences were found in serum total cholesterol, HDL-cholesterol, LDL-cholesterol, triglycerides or lipoprotein A in patients receiving letrozole or placebo [26]. Notably, in MA.17L, fasting serum lipid levels were measured in a standardized method at baseline and at regular intervals thereafter. Furthermore, the comparator in MA.17 was placebo, and this study may, therefore, more accurately reflect the true toxicity profile of letrozole.

*Quality of life*

Analysis of quality-of-life data from 3582 women in the extended adjuvant trial indicated that, compared with placebo, letrozole treatment had only minor side effects that were predictable based on its oestrogen-suppressing activity and safety profile. There were no significant differences compared with placebo in global physical or mental quality-of-life summary scores [35].

### *MA.17 re-randomisation*

The risk of late recurrences of breast cancer continues over time, and MA.17 is being extended with the aim of defining the optimal duration of letrozole therapy, determining the long-term toxicity profile, particularly in terms of bone mineral density and lipid profile, and obtaining long-term quality-of-life information. Women initially randomised to receive letrozole in the MA.17 trial who are disease-free at the completion of 4–5.5 years of extended adjuvant letrozole will be offered re-randomisation to receive either letrozole or placebo for a further 5 years. Patients will be re-randomised to the lipid and bone mineral density sub-studies and the collection of quality-of-life data will continue. The primary clinical endpoint is DFS, and secondary endpoints include the incidence of contralateral breast cancer, overall survival and quality-of-life assessments.

### *Summary of letrozole as extended adjuvant treatment*

The risk of recurrence remains significant for patients with node-positive or node-negative disease after adjuvant tamoxifen therapy. The results of MA.17 have shown that letrozole is the first agent that provides a significant benefit to patients in the extended adjuvant setting.

The MA.17 trial showed that letrozole provides a statistically significant and clinically relevant reduction in recurrence of early breast cancer in postmenopausal women, regardless of nodal status. Letrozole significantly reduced the risk of distant metastases in all patients and was associated with a statistically significant survival advantage in patients with node-positive tumours. Importantly, the side-effect profiles of letrozole and placebo were similar in this setting.

## First-line endocrine therapy for advanced breast cancer

Antioestrogen therapy with tamoxifen has been commonly used as first-line endocrine treatment for metastatic breast cancer. However, there are a number of reasons why a specific aromatase inhibitor, such as letrozole, may be preferable. Tamoxifen is routinely administered as adjuvant therapy in women with hormone receptor-positive tumours. Therefore, patients who experience relapse or progression after previous tamoxifen therapy are likely to have tumours that no longer respond to antioestrogen therapy. As aromatase inhibitors have a different mechanism of action from tamoxifen, the effectiveness of aromatase inhibitors is not likely to be diminished in some tumours that have become resistant to tamoxifen. In addition, aromatase inhibitors have a favourable side-effect profile and may offer tolerability advantages over tamoxifen.

### *Letrozole versus tamoxifen as first-line therapy*

Letrozole and tamoxifen were compared in the first-line treatment of postmenopausal women with hormone receptor-positive or -unknown locally advanced or metastatic breast cancer in a phase III trial, which remains the largest single study of its kind conducted to date [17, 18].

The aim of this double-blind, double-dummy, crossover study was to compare letrozole 2.5 mg with tamoxifen 20 mg, each administered orally once daily, as first-line treatment of locally advanced or metastatic breast cancer in postmenopausal women with ER+ and/or PgR+ or receptor-unknown tumours (Fig. 8). This multinational trial enrolled and randomised 916 patients (458 in the letrozole group and 458 in the tamoxifen group) with histologically or cytologically confirmed breast cancer and either locally advanced disease (stage IIIB), local-regionally recurrent disease not amenable to surgery or radiotherapy, or metastatic disease. Enrolment criteria required patients to have measurable or evaluable ER+ and/or PgR+ tumours or tumours with unknown status of both receptors. Patients showing progressive disease after a single regimen of cytotoxic chemotherapy for advanced disease were allowed to enrol, but prior systemic endocrine therapy for advanced disease was not permitted.

Tumour size evaluation [using Union Internationale Contre le Cancer (UICC) criteria], performance status and laboratory assessments were performed at baseline and every 3 months thereafter. Patients continued treatment

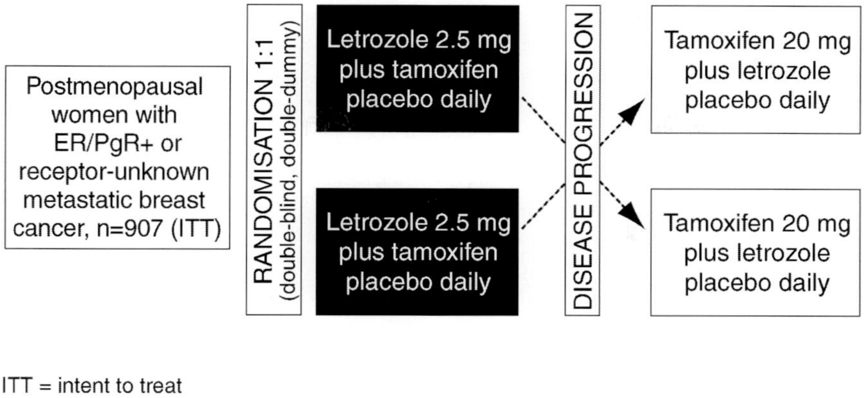


Figure 8. Design of study comparing letrozole with tamoxifen for first-line endocrine therapy in advanced breast cancer.

until development of progressive disease or discontinuation for any other reason. Following disease progression or treatment discontinuation due to an adverse event, a patient could cross over to the alternative treatment arm in a double-blind fashion, if further endocrine therapy was considered appropriate.

The primary efficacy endpoint was TTP; the main secondary endpoint was overall ORR. Additional secondary endpoints were time to treatment failure, duration of overall response, rate of clinical benefit, duration of clinical benefit and overall survival. Prior to the database being locked, analysis of survival at 6-month intervals was added as a predetermined analysis in both treatment arms. An exploratory analysis of survival, with time to death censored at crossover, was also prospectively planned to eliminate the confounding effects of the crossover on overall survival.

The intent-to-treat population comprised 453 patients in the letrozole arm and 454 in the tamoxifen arm. The study population in each treatment arm was well balanced with respect to medical history and concomitant conditions [18]. Sixty-five percent of patients in the letrozole group and 67% in the tamoxifen group had ER+ and/or PgR+ tumours. Approximately 20% of patients had received adjuvant chemotherapy: less than 20% of patients had received adjuvant antioestrogen therapy. Of those who had received prior adjuvant tamoxifen therapy, 109/167 had done so for at least 2 years. The treatment-free period prior to enrolment in this study was more than 2 years for most of these patients (126/167).

*Results of efficacy endpoints*

Results from the final analysis demonstrated a median TTP of 9.4 months for letrozole compared with 6.0 months for tamoxifen. Thus, letrozole resulted in a significant increase in the median TTP (57% or 3.4 months; $P < 0.0001$), with a hazard ratio of 0.72, and was clearly superior to tamoxifen (Tab. 4) [17].

Table 4. Summary of efficacy results from a comparative study of letrozole and tamoxifen as first-line endocrine therapy in advanced breast cancer [17]

| Endpoint | | Letrozole (n = 453) | | Tamoxifen (n = 454) | Hazard ratio (95% CI) | *P* value |
|---|---|---|---|---|---|---|
| Median TTP | | 9.4 months | | 6.0 months | 0.72 (0.62–0.83) | <0.0001 |
| Median duration of response* | | 24.7 months | | 22.9 months | 0.74 (0.54–1.01) | 0.0578 |
| | n | % (95% CI) | n | % (95% CI) | Odds ratio (95% CI) | *P* value |
| ORR (CR + PR) | 145 | 32 (28–36) | 95 | 21 (17–25) | 1.78 (1.32–2.40) | 0.0002 |
| 1-year survival | | 83 | | 75 | | 0.004 |
| 2-year survival | | 62 | | 57 | | 0.02 |

* Calculated from date of randomisation.
CI, confidence interval; CR, complete response; PR, partial response

At a median follow-up of 32 months, patients treated with letrozole were 28% less likely to progress than those treated with tamoxifen ($P < 0.0001$) (Tab. 4). Stratified multivariate analysis of TTP indicated that letrozole is consistently better than tamoxifen across relevant study subsets regardless of prior adjuvant treatment, receptor status or dominant site of metastatic disease (Fig. 9) [17]. In addition, results from the prospectively defined secondary

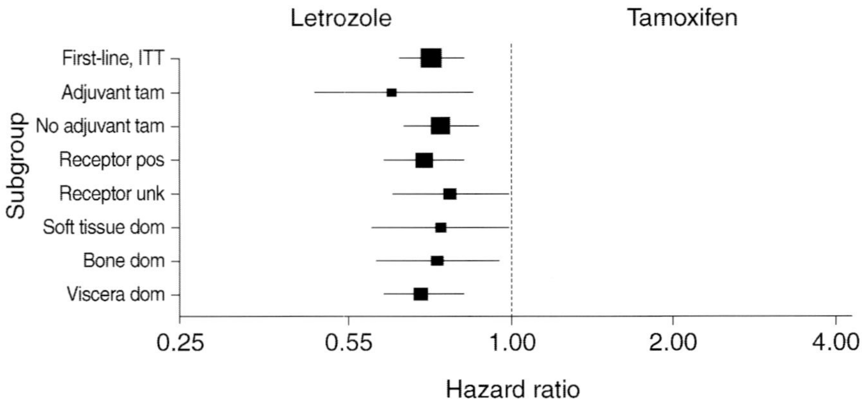


Squares donate hazard ratios, drawn proportional to the number of events, and lines represent 95% confidence intervals
Hazard ratios <1 favour letrozole
ITT = intent-to-treat; tam = tamoxifen; pos = positive; unk = unknown; dom = dominant

Figure 9. Stratified multivariate analysis shows that letrozole is better than tamoxifen in prolonging TTP, independent of prior treatment, receptor status or site of disease. Reproduced with permission [17].

endpoints of clinical benefit and time to treatment failure supported the results of the primary efficacy endpoints.

Letrozole also resulted in superior overall response rates. Patients treated with letrozole achieved a significantly greater overall ORR (32%) than those treated with tamoxifen (21%; $P = 0.0002$) as well as a higher rate of clinical benefit (50% *versus* 38%; $P = 0.0004$). Patients with hormone receptor-positive disease, previous antioestrogen therapy, and dominant site of disease in soft tissue or viscera demonstrated statistically significantly greater overall response rate with letrozole than with tamoxifen. Letrozole was significantly superior to tamoxifen in patients who had received prior adjuvant antioestrogen therapy (ORR letrozole *versus* tamoxifen 29% *versus* 8%; $P = 0.002$) [18]. In this subset of patients, using Mantel–Haenszel logistic regression analysis, the odds of response to letrozole were more than four times greater than the odds of response to tamoxifen [18].

*Crossover data and survival*

This trial was prospectively designed so that, at disease progression, patients considered appropriate for second-line endocrine therapy were permitted to crossover from letrozole to tamoxifen, or from tamoxifen to letrozole. Crossover to the alternativ arm occurred in 51% of first-line letrozole patients (median time of crossover 17 months) and 49% of those initially treated with tamoxifen (median time of crossover 13 months) [18]. Median overall survival was longer for letrozole (34 months) than for tamoxifen (30 months), but the difference was not statistically significant [18]. It was expected that crossover could have a negative impact on long-term differences between the two drugs, so prospectively planned survival analyses were performed at 6-month intervals. Significantly more patients receiving first-line letrozole than first-line tamoxifen were alive at each 6-month interval during the first 2 years of treatment (all comparisons $P < 0.025$). These results indicate the superiority of letrozole over tamoxifen in reducing the risk of death throughout the first 2 years.

Approximately 50% of patients did not crossover to the alternative treatment arm. Exploratory analysis of survival in these patients at a median of 32 months of follow-up revealed considerably longer survival in those treated with letrozole than with tamoxifen (35 months *versus* 20 months) [36]. In addition, in an analysis of all patients, censoring time to death at crossover, letrozole resulted in a 12-month survival benefit (42 months *versus* 30 months) [37].

*Time to chemotherapy*

Median time to chemotherapy was prolonged by 7 months by letrozole in comparison with tamoxifen (16.3 *versus* 9.3 months; $P = 0.005$) [17]. Thus, letrozole nearly doubled the time to chemotherapy relative to tamoxifen, sparing patients the toxicities associated with chemotherapy. Not unexpectedly, letrozole was associated with better patient performance: time to worsening of Karnofsky performance status by 20 points or more was significantly delayed with letrozole compared with tamoxifen (hazard ratio 0.62; $P = 0.001$) [17]. In addition, significantly fewer patients receiving letrozole experienced a clinically relevant deterioration in performance status compared with those receiving tamoxifen (19% *versus* 25%, odds ratio 0.69; $P = 0.02$) [17].

*Side-effect profile*

In this pivotal study, the letrozole side-effect profile was comparable with tamoxifen and was consistent with the letrozole safety profile previously reported for second-line therapy. Bone pain, hot flushes, back pain, nausea, arthralgia, dyspnoea, fatigue, coughing, constipation, chest pain, and headache were the commonly reported adverse events for letrozole and tamoxifen [18]. Discontinuations for adverse experiences occurred in 2% of patients on letrozole and in 3% of patients on tamoxifen.

### *Summary of first-line treatment with letrozole*

In conclusion, data from this first-line study of postmenopausal women with advanced breast cancer demonstrate the consistently superior efficacy of letrozole compared with tamoxifen and strongly support the use of letrozole in the first-line endocrine treatment of postmenopausal women with hormone receptor-positive or -unknown locally advanced or metastatic breast cancer. Median TTP was significantly longer in the letrozole group than in the tamoxifen group (9.4 months *versus* 6.0 months; $P < 0.0001$). Furthermore, patients treated with letrozole attained a higher overall ORR (32%) compared with those treated with tamoxifen (21%; $P = 0.0002$), as well as a higher rate of clinical benefit (50% *versus* 38%; $P = 0.0004$). Letrozole also prolonged the time to chemotherapy (16.3 *versus* 9.3 months), and delayed deterioration in performance status (54 months *versus* 43 months) compared with tamoxifen. Survival rates at 1 and 2 years were significantly greater with letrozole than tamoxifen, indicating a survival benefit with letrozole ($P = 0.004$ and $P = 0.02$, respectively). Median survival for patients who did not crossover between the treatment arms was considerably longer with letrozole than with tamoxifen (35 *versus* 20 months) and, for patients who did cross over, when data were censored at crossover, a difference in median survival was still apparent (42 *versus* 30 months).

## Second-line endocrine therapy in advanced breast cancer

Letrozole was first approved for the treatment of advanced breast cancer in postmenopausal women with disease progression following antioestrogen therapy. The efficacy of letrozole as endocrine therapy for advanced breast cancer in postmenopausal women previously treated with antioestrogens has been demonstrated in pivotal clinical trials that compared letrozole with the progestin megestrol acetate [19, 38] or with the aromatase inhibitor aminoglutethimide [13]. A further study directly compared the non-steroidal aromatase inhibitors, letrozole and anastrozole [14].

*Comparison with megestrol acetate*

The antitumour efficacy of three treatment regimens: letrozole 0.5 mg, letrozole 2.5 mg, and megestrol acetate 160 mg, each administered orally once daily, were initially compared in a double-blind, randomised, multicentre trial that recruited 551 patients with advanced breast cancer [19]. Patients were postmenopausal women with locally advanced, locally recurrent, or metastatic breast cancer who had objective evidence of disease progression following antioestrogen treatment for either metastatic disease or adjuvant treatment of localised breast cancer, ER+ and/or PgR+ status (57%) or receptor status unknown (43%), and measurable or evaluable disease.

The primary efficacy endpoint was overall response rate (complete plus partial responses). Secondary efficacy endpoints were duration of response, TTP, and overall survival. All available data were analysed for tumour response and safety variables for up to 33 months of follow-up and for survival for up to 45 months. All analyses were conducted using an intent-to-treat approach.

Another double-blind, randomised, multicentre study compared two doses of letrozole, 0.5 mg/day and 2.5 mg/day, and megestrol acetate, 40 mg q.d.s, in 602 postmenopausal women with advanced or metastatic breast cancer previously treated with antioestrogens [38]. Tumours were ER+ or PgR+ or of unknown receptor status. The primary efficacy endpoint was confirmed ORR.

*Response rates*

In the first study, letrozole 2.5 mg achieved an overall response rate of 23.6%, compared with 12.8% with letrozole 0.5 mg ($P = 0.004$) and 16.4% with megestrol acetate ($P = 0.04$) (Tab. 5) [19]. The likelihood of achieving a response for letrozole 2.5 mg was 58% higher than for megestrol acetate. Subgroup analyses were performed to examine the effect of other prognostic factors on outcome [19, 22]. Among patients who had not responded to initial antioestrogen therapy (refractory), 29% achieved an objective response with letrozole 2.5 mg, compared with 15% with megestrol acetate. There was a trend towards higher response rates for all disease sites (soft tissue, bone, viscera) with letrozole (Tab. 5).

The duration of response (Kaplan-Meier estimate) was significantly longer with letrozole 2.5 mg (more than 33 months, median not reached at time of analysis) than with megestrol acetate (median 17.9 months, $P = 0.02$). Although the median TTP values with letrozole 2.5 mg and megestrol acetate were similar (5.6 *versus* 5.5 months, respectively), patients receiving letrozole 2.5 mg had

Table 5. Comparative efficacy of letrozole and megestrol acetate in women with metastatic breast cancer after antioestrogen failure [19]

| Primary and secondary endpoints | Letrozole 0.5 mg (n = 188) | Letrozole 2.5 mg (n = 174) | MA 160 mg (n = 189) |
|---|---|---|---|
| Objective tumour response | 24 (12.8%) | 41 (23.6%) | 31 (16.4%) |
| Median duration of response (months) | 18.2 | >33 (not reached) | 17.9 |
| Median TTP (months) | 5.1 | 5.6 | 5.5 |
| **Objective response rates by disease site** | | **Letrozole 2.5 mg %** | **MA 160 mg %** |
| Soft tissue metastasis only | | 47.9 | 40 |
| Bone ± soft tissue | | 15 | 10 |
| Viscera ± bone ± soft tissue | | 16 | 8 |

MA = megestrol acetate.

a 23% lower risk of disease progression than those receiving megestrol acetate ($P = 0.03$). The difference in median overall survival in the two groups was not statistically significant: 24 months in those receiving letrozole 2.5 mg compared with 21.6 months in the megestrol acetate group [19].

This first study demonstrated the clinical efficacy of once-daily letrozole 2.5 mg for the treatment of advanced breast cancer in postmenopausal women with disease progression following antioestrogen therapy.

In the second study, no significant differences were found between either of the two letrozole treatment groups and megestrol acetate group in terms of ORR [38]. However, patients treated with letrozole 0.5 mg had a significantly lower risk of disease progression ($P = 0.044$) and a significantly reduced risk of treatment failure ($P = 0.018$) compared with patients treated with megestrol acetate [38]. Although the results of this study do not replicate the statistically significant superiority of letrozole 2.5 mg *versus* megestrol acetate, letrozole 0.5 mg showed clinical benefit, providing further evidence of the activity of letrozole in patients with advanced breast cancer who have experienced progression despite antioestrogen therapy. Heterogeneity among trials is to be expected in this poor-prognosis patient population and may be attributable to variation in patient characteristics.

### *Comparison with aminoglutethimide*

The antitumour efficacy of letrozole and aminoglutethimide was compared in an open-label, randomised, multinational, multicentre trial with three treatment arms: letrozole 0.5 mg and letrozole 2.5 mg, both administered once daily, and aminoglutethimide 250 mg administered twice daily with corticosteroid supplementation (hydrocortisone 30 mg or cortisone acetate 37.5 mg daily) [13].

The study recruited 555 postmenopausal women with hormone receptor-positive or -unknown advanced breast cancer with objective evidence of relapse during or within 1 year following adjuvant antioestrogen treatment, or disease progression during antioestrogen treatment for advanced disease. Across the three groups, 50–60% of patients were hormone receptor-positive.

The primary efficacy endpoint was ORR, evaluated according to UICC criteria. Secondary efficacy endpoints were duration of response, TTP, and survival. All available data were analysed 9 months after the last patient was enrolled and all analyses were based on the intent-to-treat approach.

### *Disease control*

Whereas there was a trend towards improved response with letrozole 2.5 mg compared with aminoglutethimide ($P = 0.06$), overall response rates were not statistically significantly different between the two treatment arms (19.5% *versus* 12.4%, respectively) or between letrozole 0.5 mg and 2.5 mg (Tab. 6) [13]. Median duration of response was longer for patients treated with letrozole 2.5 mg than with aminoglutethimide, but the difference was not statistically significant (24 months *versus* 15 months; Table 6) [13]. Median TTP was 3.4 months for patients treated with letrozole 2.5 mg compared with 3.2 months for those treated with aminoglutethimide (Tab. 6) [13]. Cox regression analysis over a follow-up period of 27 months indicated significantly longer TTP with letrozole 2.5 mg than with aminoglutethimide ($P = 0.008$) [13].

Median survival was also longer for patients treated with letrozole 2.5 mg (28 months) than aminoglutethimide (20 months; Table 6). Cox regression analysis over a follow-up period of 27 months indicated that the longer sur-

Table 6. Efficacy outcomes of letrozole and aminoglutethimide in postmenopausal women with advanced breast cancer [13]

| | Letrozole 2.5 mg | Letrozole 0.5 mg | AG |
|---|---|---|---|
| ORR (%) | 19.5 | 16.7 | 12.4 |
| Clinical benefit (%) | 36 | 33 | 29 |
| MDR (months) | 24 | 21 | 15 |
| MDCB (months) | 21 | 18 | 14 |
| Median TTP (months) | 3.4 | 3.3 | 3.2 |
| Median overall survival (months) | 28 | 21 | 20 |

AG = aminoglutethimide; MDR = median duration of response; MDCB = median duration of clinical benefit.

vival with letrozole 2.5 mg compared with aminoglutethimide was statistically significant ($P = 0.002$) [13].

Treatment-related adverse events occurred in fewer patients receiving letrozole 2.5 mg (33%) than in those receiving aminoglutethimide (46%). Transient nausea and rash were the most commonly seen adverse events, and the incidence of the latter was higher for patients receiving aminoglutethimide (11%) than for those receiving letrozole 2.5 mg (3%) [13].

### *Comparison with anastrozole*

In a direct comparison, the ORR to letrozole proved superior to that of anastrozole in an open-label, randomised, multicentre trial in patients with hormone receptor-positive or -unknown metastatic breast cancer who had progressed during or within 1 year of first-line antioestrogen therapy for advanced disease [14]. The study recruited 713 women with metastatic breast cancer after failure on antioestrogen therapy. Hormone receptor status was positive in 48% and unknown in 52% of the patient population. Patients with documented ER/PgR-negative status were excluded from this trial. Visceral disease was present in 52% of patients and 24% had bone-dominant disease.

The study was powered to detect a 30% difference (hazard ratio 1.3) between letrozole and anastrozole in the primary endpoint, TTP. Secondary endpoints included ORR, duration of response, clinical benefit, duration of clinical benefit, time to treatment failure, and survival.

Patients treated with letrozole 2.5 mg were 50% more likely to respond to therapy than those treated with anastrozole 1 mg: an objective response was observed in 19% of patients in the letrozole arm compared with 12% in the anastrozole arm ($P = 0.013$) [14], response rates that are consistent with previous findings with these agents in the second-line setting [37]. More patients with soft tissue- or visceral-dominant disease responded to letrozole (37% and 14%, respectively) than to anastrozole (19% and 10%, respectively) [14]. The outcome for patients with visceral disease treated with letrozole was consistent with results obtained in other second-line clinical trials, which show a 15–17% response rate in this patient population. When patients were stratified on the basis of receptor status, the superior ORR to letrozole remained significant only in patients with unknown-receptor status.

There was no significant difference between letrozole and anastrozole with regard to either the primary endpoint (TTP) or overall survival [14]. Although a significant difference was seen in ORR between the letrozole and anastrozole arms in this study, when patients were stratified on the basis of receptor status an improvement in ORR was only seen in those with unknown receptor status. This study was undoubtedly underpowered and is open to criticism on the basis of the open-label design. Although the results of this direct comparative study provide some support for the clinical superiority of letrozole over anastrozole, they are not definitive.

### *Summary of letrozole in second-line clinical trials*

The studies comparing letrozole with megestrol acetate and aminoglutethimide demonstrated that letrozole has significant efficacy and tolerability advantages over both agents for the treatment of advanced breast cancer in postmenopausal women with disease progression following antioestrogen therapy.

In a comparative trial of letrozole and anastrozole, letrozole achieved a significantly higher response rate than anastrozole in patients with advanced breast cancer that had progressed following antioestrogen therapy [14]. The results of this direct comparison between letrozole and anastrozole may reflect the greater aromatase inhibition and oestrogen suppression that has been demonstrated for letrozole compared with anastrozole [20].

### *Further developments in advanced disease*

#### *FRAGRANCE trial*

The Femara Reanalysed through Genomics for Response Assessment, Calibration and Empowerment (FRAGRANCE) trial has the objective of defining the efficacy of letrozole with or without the antiproliferative macrolide RAD001 for tumour shrinkage before surgery and to identify factors predictive of response to neoadjuvant letrozole, based on specific characteristics of the tumour.

Other developments include clinical trials with the combination of letrozole and the farnesyltransferase inhibitors erlotinib (OSI-774) or tipifarnib (R115777).

#### *Erb-B2 (HER2/neu)-overexpressing breast cancer*

Several studies have linked Erb-B1 (epidermal growth factor receptor) and Erb-B2 (HER2/*neu*) expression in breast cancer to tamoxifen resistance [39–45]. Preclinical modelling is consistent with the conclusion that ER+ and HER2/*neu*+ tumours are oestrogen-dependent [46]. It has been shown that MCF-7 breast cancer cells transfected with a HER2/*neu* expression vector grow rapidly as xenografts in nude mice supplemented with oestrogen. When oestrogen supplementation is stopped and tamoxifen treatment started, control HER2/*neu*– xenografts stop growing and regress, whereas HER2/*neu*+ xenografts continue to grow in the presence of tamoxifen [46]. A possible molecular explanation for this finding was provided by a recent observation that a downstream mediator of Erb-B1/2 signalling, MEKK1, activates the ER and stimulates the agonist activity of tamoxifen [23]. The Erb-B1/2 tamoxifen resistance pathway may be circumvented by letrozole. As letrozole has no agonist-like activity for the ER, MEKK1-mediated activation does not occur, which precludes receptor dimerization and abrogates ER-mediated transcription and downstream signalling. Hence, in this setting, the ER is not a productive target for Erb-B1/2-activated protein kinases [23].

HER2/*neu* gene amplification or protein overexpression is present in 20–30% of primary breast cancers [47–50], and a difference in activity between letrozole and tamoxifen in HER2/*neu*+ tumours would have important implications for the use of hormonal therapies in early-stage and metastatic breast cancer.

*Letrozole compared with tamoxifen*

The P024 study compared letrozole with tamoxifen as preoperative therapy in postmenopausal women with ER+ and/or PgR+ breast cancer who were not eligible for breast-conserving surgery [15] and the trial design, patient characteristics and clinical outcomes have been described in detail earlier in the section on 'Primary systemic therapy in early breast cancer'.

This study also provided an opportunity to investigate the biological basis for the response to letrozole and tamoxifen. A prospective analysis was undertaken to explore relationships between ER and/or PgR expression levels and response rates, as well as between Erb-B1 and HER2/*neu* expression and response rates. Tumour samples were analysed for ER, PgR, HER2/*neu*, and Erb-B1 expression using immunohistochemistry. All study analyses were blinded with respect to clinical outcomes, patient identity, and drug assignment.

This biomarker study revealed possible molecular explanations for the superiority of letrozole over tamoxifen. For example, in tumours that were both Erb-B1+ and/or HER2/*neu*+ and ER+, overexpression of Erb-B1 and/or HER2/*neu* was a significant predictive marker for selective response to treatment with letrozole but not tamoxifen. Although this subgroup was small ($n = 36$), the difference was highly significant, with 15/17 (88%) patients responding to letrozole, while only 4/19 (21%) responded to tamoxifen ($P = 0.0004$) [23].

They also suggest that the Erb-family receptor HER2/*neu* is associated with tamoxifen resistance. HER2/*neu* is overexpressed in 20–30% of primary breast cancers, and letrozole appears superior to tamoxifen in ER+ and Erb-B1+ and/or HER2/*neu*+ primary breast cancer.

A further study investigated the interaction between HER2 status and response to neoadjuvant letrozole [51]. The study recruited 172 postmenopausal women with large operable or locally advanced ER-rich (Allred score ≥5) tumours into a prospective audit assessing response to 3 months of neoadjuvant letrozole 2.5 mg/day. Response rate and reduction in tumour area and volume in HER2 positive (3+ or 2+ and FISH positive) tumours were compared with tumours classified as HER2 negative. No significant differences were found between tumour responses, in terms of clinical area or volume and ultrasound area or volume, in the groups. This study found that the response to neoadjuvant letrozole in postmenopausal women with large operable or locally-advanced ER- rich breast cancer is not related to HER2 status.

## References


1 Imaginis Corporation. At: http://www.imaginis.com/breasthealth/ statistics.asp. Accessed May 31, 2005

2 Ferlay J, Bray F, Pisani P, Parkin DM (2001) GLOBOCAN 2000. IARC Press, Lyon, France

3 Early Breast Cancer Trialists' Collaborative Group (1998) Tamoxifen for early breast cancer: an overview of the randomized trials. *Lancet* 351: 1451–1467

4 Dowsett M (1992) Rationale for the endocrine treatment of breast cancer. In: Dowsett M (ed.): *Endocrine Aspects of Breast Cancer.* Parthenon Publishing Group, Camforth, England, 11–24

5 Buzdar AU (1990) Current status of endocrine treatment of carcinoma of the breast. *Semin Surg Oncol* 6: 77–82

6 Masamura S, Adlercreutz H, Harvey H, Lipton A, Demers LM, Santen RJ, Santner SJ (1995) Aromatase inhibitor development for treatment of breast cancer. *Breast Cancer Res Treat* 33: 19–26

7 Lønning P, Pfister C, Martoni A, Zamagni C (2003) Pharmacokinetics of third-generation aromatase inhibitors. *Semin Oncol* 30 (Suppl 14): 23–32

8 Grodin JM, Siiteri PK, MacDonald PC (1973) Source of estrogen production in postmenopausal women. *J Clin Endocrinol Metab* 36: 207–214

9 Goldhirsch A, Gelber RD (1996) Endocrine therapies in breast cancer. *Semin Oncol* 23: 494–505

10 Bhatnagar AS, Hausler A, Schieweck K, Lang M, Bowman R (1990) Highly selective inhibition of estrogen biosynthesis by CGS 20267, a new non-steroidal aromatase inhibitor. *J Steroid Biochem Mol Biol* 37: 1021–1027

11 Goss PE, Ingle JN, Martino S, Robert NJ, Muss HB, Piccart MJ, Castiglione M, Tu D, Shepherd L, Pritchard KI et al. (2003) A randomized trial of letrozole in postmenopausal women after five years of tamoxifen therapy for early-stage breast cancer. *N Engl J Med* 349: 793–802

12 Goss PE, Ingle JN, Martino S, Robert NJ, Muss HB, Piccart MJ et al. (2004) Updated analysis of the NCIC CTG MA.17 randomized placebo controlled trial of letrozole after five years of tamoxifen in postmenopausal women with early stage breast cancer. *Proc Am Soc Clin Oncol* 22: 87 (abstract 847)

13 Gershanovich M, Chaudri HA, Campos D, Lurie H, Bonaventura A, Jeffrey M, Buzzi F, Bodrogi I, Ludwig H, Reichardt P et al. for the Letrozole International Trial Group (AR/BC3) (1998) Letrozole, a new oral aromatase inhibitor: randomised trial comparing 2.5 mg daily, 0.5 mg daily and aminoglutethimide in postmenopausal women with advanced breast cancer. *Ann Oncol* 9: 639–645

14 Rose C, Vtoraya O, Pluzanska A, Davidson N, Gershanovich M, Thomas R, Johnson S, Caicedo JJ, Gervasio H, Manikhas G et al. (2003) An open randomised trial of second-line endocrine therapy in advanced breast cancer: comparison of the aromatase inhibitors letrozole and anastrozole. *Eur J Cancer* 39: 2318–2327

15 Eiermann W, Paepke S, Appfelstaedt J, Llombart-Cussac A, Eremin J, Vinholes J, Mauriac L, Ellis M, Lassus M, Chaudri-Ross HA et al. (2001) Preoperative treatment of postmenopausal breast cancer patients with letrozole: A randomized, double-blind, multicenter study. *Ann Oncol* 12: 1527–1532

16 [http://www.ibcsg.org/public/documents/pdf/divers/BIG_1-98_ASCO_2005_Final.pdf]. Accessed June 2, 2005

17 Mouridsen H, Gershanovich M, Sun Y, Pérez-Carrión R, Boni C, Monnier A, Apffelstaedt J, Smith R, Sleeboom HP, Jänicke F et al. (2003) Phase III study of letrozole *versus* tamoxifen as first-line therapy of advanced breast cancer in postmenopausal women: analysis of survival and update of efficacy from the International Letrozole Group. *J Clin Oncol* 21: 2101–2109

18 Mouridsen H, Gershanovich M, Sun Y, Pérez-Carrión R, Boni C, Monnier A, Apffelstaedt J, Smith R, Sleeboom HP, Jänicke F et al. (2001) Superior efficacy of letrozole *versus* tamoxifen as first-line therapy for postmenopausal women with advanced breast cancer: results of a phase III study of the International Letrozole Group. *J Clin Oncol* 19: 2596–2606

19 Dombernowsky P, Smith I, Falkson G, Leonard R, Panasci L, Bellmunt J, Bezwoda W, Gardin G, Gudgeon A, Morgan M et al. (1998) Letrozole, a new oral aromatase inhibitor for advanced breast cancer: double-blind randomized trial showing a dose effect and improved efficacy and tolerability compared with megestrol acetate. *J Clin Oncol* 16: 453–461

20 Geisler J, Haynes B, Anker G, Dowsett M, Lönning PE (2002) Influence of letrozole and anastrozole on total body aromatization and plasma estrogen levels in postmenopausal breast cancer

patients evaluated in a randomized, cross-over study. *J Clin Oncol* 20: 751–757
21 Dixon JM, Love CD, Bellamy CO, Cameron DA, Leonard RC, Smith H, Miller WR (2001) Letrozole as primary medical therapy for locally advanced and large operable breast cancer. *Breast Cancer Res Treat* 66: 191–199
22 Data on file. Novartis Pharmaceuticals Corporation
23 Ellis MJ, Coop A, Singh B, Mauriac L, Lombert-Cussac A, Jänicke F, Miller WR, Evans DB, Dugan M, Brady C et al. (2001) Letrozole is more effective neoadjuvant endocrine therapy than tamoxifen for ErbB-1 and/or ErbB-2-positive, estrogen receptor-positive primary breast cancer: evidence from a phase III randomized trial. *J Clin Oncol* 19: 3808–3816
24 Allred DC, Harvey JM, Bernardo M, Clark GM (1998) Prognostic and predictive factors in breast cancer by immunohistochemical analysis. *Mod Pathol* 11: 155–168
25 Renshaw L, Murray J, Young O, Cameron D, Miller WR, Dixon JM (2004) Is there an optimal duration of neoadjuvant letrozole therapy. *Breast Cancer Research and Treatment*; *in press*
26 Wasan KM, Goss PE, Pritchard PH, Shepherd L, Palmer MJ, Liu S, Tu D, Ingle JN, Heath M, Deangelis D, Perez EA (2005) The influence of letrozole on serum lipid concentrations in postmenopausal women with primary breast cancer who have completed 5 years of adjuvant tamoxifen (NCIC CTG MA.17L). *Ann Oncol* 16: 707–715
27 Howell A, Cuzick J, Baum M, Buzdar A, Dowsett M, Forbes JF, Hoctin-Boes G, Houghton J, Locker GY, Tobias JS; ATAC Trialists' Group (2005) Results of the ATAC (Arimidex, Tamoxifen, Alone or in Combination) trial after completion of 5 years' adjuvant treatment for breast cancer. *Lancet* 365: 60–62
28 Coombes RC, Hall E, Gibson LJ, Paridaens R, Jassem J, Delozier T, Jones SE, Alvarez I, Bertelli G, Ortmann O et al. (2004) A randomized trial of exemestane after two to three years of tamoxifen therapy in postmenopausal women with primary breast cancer. *N Engl J Med* 350: 1081–1092
29 Hillner BE, Ingle JN, Chlebowski RT, Gralow J, Yee GC, Janjan NA, Cauley JA, Blumenstein BA, Albain KS, Brown S (2003) American Society of Clinical Oncology 2003 update on the role of bisphosphonates and bone health issues in women with breast cancer. *J Clin Oncol* 21: 4042–4057
30 Brufsky A, Harker G, Beck T, Carroll R, Tan-Chiu E, Seidler C, Lacema L, Thomas E, Perez E; Z-FAST Study Group (2004) Zoledronic acid (ZA) for prevention of cancer treatment-induced bone loss (CTIBL) in postmenopausal women (PMW) with early breast cancer (BCa) receiving adjuvant letrozole (Let): Preliminary results of the Z-FAST trial. *Breast Cancer Res Treat* 88: S233(abstract 6038)
31 Saphner T, Tormey DC, Gray R (1996) Annual hazard rates of recurrence for breast cancer after primary therapy. *J Clin Oncol* 14: 2738–2746
32 Hortobagyi GN, Kau S-W, Buzdar AU, Theriault RL, Booser DJ, Gwyn K, Valero V (2004) What is the prognosis of patients with operable breast cancer (BC) five years after diagnosis? *Proc Am Soc Clin Oncol* 22: (abstract 585)
33 Fisher B, Dignam J, Bryant J, Wolmark N (2001) Five *versus* more than five years of tamoxifen for lymph node-negative breast cancer: updated findings from the National Surgical Adjuvant Breast and Bowel Project B-14 randomized trial. *J Natl Cancer Inst* 93: 684–690
34 Perez EA, Josse RG, Pritchard KI, Ingle JN, Martino S, Findlay BP, Shenkier TN, Tozer RG, Palmer MJ, Shepherd LE, Tu D, Goss PE (2004) Effect of letrozole *versus* placebo on bone mineral density in women completing 5 years (yrs) of adjuvant tamoxifen: NCIC CTG MA.17B. *Breast Cancer Res Treat* 88: S36 (abstract 404)
35 Whelan T, Goss P, Ingle J, Pater J, Shepherd L, Palmer D, Tu D, Robert N, Martino S, Muss H et al. (2004) Assessment of quality of life (QOL) in MA.17, a randomized placebo-controlled trial of letrozole in postmenopausal women following five years of tamoxifen. *Proc Am Soc Clin Oncol* 23: 6 (abstract 517)
36 Femara Prescribing Information (2003) East Hanover, NJ: Novartis Pharmaceuticals Corporation
37 Mouridsen HT, Rose C, Brodie AH, Smith IE (2003) Challenges in the endocrine management of breast cancer. *Breast* 12 (Suppl 2): S2–S19
38 Buzdar A, Douma J, Davidson N, Elledge R, Morgan M, Smith R, Porter L, Nabholtz J, Xiang X, Brady C (2001) Phase III, multicenter, double-blind, randomized study of letrozole, an aromatase inhibitor, for advanced breast cancer *versus* megestrol acetate. *J Clin Oncol* 19: 3357–3366
39 Wright C, Nicholson S, Angus B, Sainsbury JR, Farndon J, Cairns J, Harris AL, Horne CH (1992) Relationship between *c-erb*B-2 protein product expression and response to endocrine therapy in advanced breast cancer. *Br J Cancer* 65: 118–121
40 Yamauchi H, O'Neill A, Gelman R, Carney W, Tenney DY, Hosch S, Hayes DF (1997) Prediction

of response to antiestrogen therapy in advanced breast cancer patients by pre-treatment circulating levels of extracellular domain of the HER-2/c-neu protein. *J Clin Oncol* 15: 2518–2525

41 Berns EM, Foekens JA, van Staveren IL, van Putten WL, de Koning HY, portengen H, Klijn JO (1995) Oncogene amplification and prognosis in breast cancer: Relationship with systemic treatment. *Gene* 159: 11–18

42 Archer SG, Eliopoulos A, Spandidos D, Barnes D, Ellis IO, Blamey RW, Nicholson RI, Robertson JF (1995) Expression of ras p21, p53 and c-erbB-2 in advanced breast cancer and response to first line hormonal therapy. *Br J Cancer* 72: 1259–1266

43 Newby JC, Johnston SR, Smith IE, Dowsett M (1997) Expression of epidermal growth factor receptor and c-erb-B2 during the development of tamoxifen resistance in human breast cancer. *Clin Cancer Res* 3: 1643–1651

44 Carlomagno C, Perrone F, Gallo C, De Laurentiis M, Lauria R, Morabito A, Pettinato G, Panico L, D'Antonio A, Bianco AR, De Placido S (1996) c-erb B2 overexpression decreases the benefit of adjuvant tamoxifen in early-stage breast cancer without axillary lymph node metastases. *J Clin Oncol* 14: 2702–2708

45 Lipton A, Ali SM, Leitzel K, Chinchilli V, Engle L, Demers L, Brady C, Carney W, Cook G et al. (2000) Elevated serum Her-2/neu predicts decreased response to hormone therapy in metastatic breast cancer. *Proc Am Soc Clin Oncol* 19: 1a (abstract 274)

46 Benz CC, Scott GK, Sarup JC et al. (1993) Estrogen-dependent, tamoxifen-resistant tumorigenic growth of MCF-7 cells transfected with HER2/neu. *Breast Cancer Res Treat* 24: 85–95

47 Lee H, Jiang F, Wang Q, Nicosia SV, Yang J, Su B, Bai W (2000) MEKK1 activation of human oestrogen receptor alpha and stimulation of the agonistic activity of 4-hydroxytamoxifen in endometrial and ovarian cancer cells. *Mol Endocrinol* 14: 1882–1896

48 Lupu R, Lippman ME (1993) The role of erbB2 signal transduction pathways in human breast cancer. *Breast Cancer Res Treat* 27: 83–93

49 Slamon DJ, Godolphin W, Jones LA, Holt JA, Wong SG, Keith DE, Levin WJ, Stuart SG, Udove J, Ullrich A et al. (1989) Studies of the HER-2/*neu* proto-oncogene in human breast and ovarian cancer. *Science* 244: 707–712

50 Clark GM, McGuire WL (1991) Follow-up study of HER-2/*neu* amplification in primary breast cancer. *Cancer Res* 51: 944–948

51 Young O, Murray J, Renshaw L, Evans D, Cameron D, Dowsett M, Miller WR, Dixon JM (2004) Her2 status does not influence response to neoadjuvant letrozole. *Breast Cancer Research and Treatment* 88(Suppl 1): S38

Aromatase Inhibitors
Edited by B.J.A. Furr



# Clinical studies with anastrozole

Anthony Howell[1] and Alan Wakeling[2]

*1CRUK Department of Medical Oncology, Christie Hospital NHS Trust, Manchester, UK*
*[2] Department of Cancer and Infection Research, AstraZeneca Pharmaceuticals, Macclesfield, UK*

## Introduction

Estrogens play a dominant role in controlling the growth of many breast cancers [1, 2]. The ovaries are the primary source of estrogen in premenopausal women but ovarian estrogen production diminishes with age. In postmenopausal women, estrogens are synthesized by aromatization of androgen precursors in the skin, muscle, adipose and breast tissue, including malignant breast tumors [3]. Inhibition of estrogen action is achieved by blocking the estrogen receptor (ER) with antiestrogens such as tamoxifen, by ovarian ablation using surgery, radiotherapy or luteinizing hormone-releasing hormone analogs such as goserelin, or, in postmenopausal women, by blocking estrogen production by inhibiting aromatase activity [4].

Aromatase catalyzes the conversion of androstenedione and testosterone into estrone and estradiol, respectively [5]. Aromatase inhibitors (AIs) suppress estrogen production by inhibiting the final step in estrogen synthesis catalyzed by the cytochrome P450 (CYP) enzyme complex aromatase. The AI first used in the clinic, aminoglutethimide, was introduced approximately 25 years ago for the second-line treatment of advanced breast cancer in postmenopausal women [6]. Despite proving clinically effective, aminoglutethimide also inhibited the synthesis of adrenal steroids, and required concomitant administration of hydrocortisone [7–9]. A second-generation, parentally administered AI, formestane, was more potent and selective for aromatase than aminoglutethimide [10, 11]; however, its use was limited by a high incidence of injection-site reactions [12]. The search for more potent, selective and well-tolerated AIs led to the discovery of anastrozole (Arimidex), the first third-generation AI to enter into clinical trials. Since its first launch in 1995, anastrozole has received approval in many countries for use in postmenopausal women with advanced, ER-positive breast cancer (>90 countries) and as an adjuvant therapy for early breast cancer (68 countries). Anastrozole and other AIs are increasingly the treatment of choice for postmenopausal women with breast cancer because they are more effective than tamoxifen [13]. This chapter summarizes the preclinical and clinical pharmacology of anastrozole and describes its current use in breast cancer therapy.

## Preclinical pharmacology of anastrozole

### *Aromatase inhibition*

Anastrozole, an achiral, benzyl triazole derivative (2,2'[5-(1H-1,2,4-triazol-1-ylmethyl)-1,3-phenylene]bis-2-methylproprionitrile; Fig. 1) is a non-steroidal inhibitor of aromatase. Non-steroidal AIs bind reversibly to the haem group of the aromatase enzyme via a basic nitrogen atom, which is on the triazole group of anastrozole. In early preclinical studies, the potency of anastrozole was assessed using human placental microsomal aromatase preparations [14]. In this *in vitro* system, anastrozole was a potent inhibitor of human placental aromatase (200 times as potent as aminoglutethimide and twice as potent as 4-hydroxyandrostenedione), with an $IC_{50}$ of 15 nM.

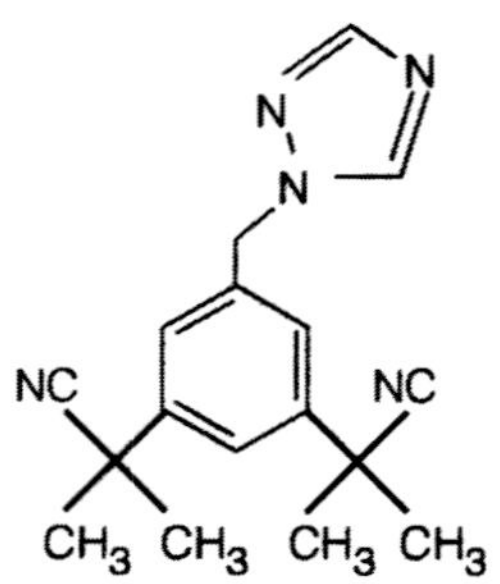


Figure 1. Structure of anastrozole.

Preclinical studies were extended to include *in vivo* functional testing in animals [14]. In adult female rats a single oral dose of anastrozole (0.1 mg/kg) given on day 2 or 3 of the estrous cycle blocked ovulation. Similarly, at the same daily dosage (0.1 mg/kg) anastrozole inhibited androstenedione-induced uterine hypertrophy in sexually immature rats. In addition, inhibition of peripheral aromatase activity was observed in male pigtailed monkeys, with twice-daily oral treatment with ≥0.1 mg/kg doses of anastrozole reducing circulating estradiol concentrations by 50–60%. Therefore, in animals, a dose of approximately 0.1 mg/kg anastrozole effectively inhibits aromatase activity.

### *Enzyme selectivity: interactions with other CYP enzymes*

*In vitro* and *in vivo* preclinical studies were used to assess the selectivity of anastrozole for aromatase compared with inhibition of other CYP enzymes responsible for steroid biosynthetic pathways. Anastrozole did not substantially inhibit cholesterol biosynthesis *in vitro* or alter plasma cholesterol concentrations *in vivo* [15]. In addition, anastrozole did not interfere with cholesterol side-chain cleavage (no adrenal hypertrophy), affect plasma aldosterone levels (indicating no effect on 18-hydroxylase activity) or alter sodium and potassium excretion [14, 15]. Although anastrozole was a comparatively weak inhibitor of bovine adrenal 11β-hyroxylase *in vitro*, it had no detectable effect on plasma 11-deoxycorticosterone concentrations in a range of animal models [14, 15].

To investigate the potential for clinically significant interactions with other CYP-metabolized drugs, the inhibitory potential of anastrozole on a range of human liver CYP isoforms (CYP1A2, 2A6, 2C9, 2D6 and 3A) was examined using a well-validated *in vitro* system [16]. At concentrations <500 μM anastrozole did not inhibit CYP2A6 and CYP2D6 activities, but CYP1A2, 2C9 and 3A were inhibited with $K_i$ values of 8, 10 and 10 μM, respectively. However, these concentrations of anastrozole are approximately 30-fold higher than the average steady-state $C_{max}$ concentrations in patients chronically administered anastrozole 1 mg/day. This suggests that anastrozole is unlikely to cause clinically significant interactions with other CYP-metabolizing drugs at therapeutic concentrations.

### *General pharmacological activity*

Preclinical studies demonstrated that anastrozole had no estrogenic or antiestrogenic activity, no androgenic activity, was not progestogenic and had no glucocorticoid/antiglucorticoid activity [14]. In addition, anastrozole alone or in combination with – and following – tamoxifen treatment, did not affect lipid metabolism (serum total cholesterol, triglycerides and lipoprotein lipase activity) in ovariectomized female rats [17].

### *Antitumor effects in model systems*

A xenograft model system using aromatase-transfected human MCF-7 breast cancer cells in ovariectomized nude mice has been developed to simulate the postmenopausal breast cancer patient with ER-positive tumors [18]. Using this system anastrozole has been shown to be a potent suppressor of tumor growth. Administration of anastrozole (10 μg/day) to tumor-bearing mice for 28 days prevented tumor growth [19]. During the same period, anastrozole at a dose of 60 μg/day reduced tumor volume by approximately 20% from initial size. In this xenograft tumor model anastrozole was more effective than tamoxifen in preventing tumor growth. In addition, the effect of inhibiting both estrogen action and estrogen synthesis by combining tamoxifen with anastrozole was investigated in the tumor model [19]. The combination of anastrozole and tamoxifen tended to be less effective than anastrozole alone, but this difference was not statistically significant.

*Pharmacokinetics*

In studies with radiolabelled compound, anastrozole was absorbed rapidly and almost completely after oral administration to animals [20]. Although the clearance half-life after a single dose of anastrozole in rats and monkeys was approximately 7–8 h, pharmacokinetic data from dogs (clearance half-life of around 16 h) suggested that once-daily dosing of anastrozole in humans would be feasible.

## Clinical pharmacology

*Aromatase inhibition*

*Suppression of circulating estrogen levels*
The effect of anastrozole on serum estradiol levels in postmenopausal women was investigated in three phase 1 studies [15]. In one study, postmenopausal female volunteers received 14 once-daily oral doses of 0.5 or 1.0 mg of anastrozole. The second study was a double-blind crossover trial in healthy postmenopausal volunteers that evaluated a 3 mg daily dose of anastrozole over a period of 10 days. The third study, involving postmenopausal women with advanced breast cancer, investigated the effects of anastrozole 5 mg/day for 14 days followed by administration of the drug at 10 mg/day for another 14 days. In all three studies, for each of the doses evaluated, maximal suppression of plasma estradiol occurred after 3–4 days of treatment, with reductions in estradiol of approximately 80% of baseline or to the limit of detection of the assay [21, 22]. A dose of 1 mg of anastrozole per day was the smallest amount required for complete suppression of estradiol and is the approved dose of the drug.

The long-term effects that a daily dose of 1 mg of anastrozole has on plasma estrogen levels were determined in a randomized, open-label, parallel-group trial comparing oral anastrozole with intramuscular formestane (250 mg, once every 2 weeks) in postmenopausal women with advanced breast cancer [23]. Anastrozole (1 mg, by mouth, once daily) provided constant and reliable estradiol suppression over a 4-week period (Fig. 2). Anastrozole produced greater suppression of serum estradiol, estrone and estrone sulfate levels compared with formestane ($P < 0.005$ in all cases). These data show that in postmenopausal women with advanced breast cancer anastrozole provides a consistent and significantly greater estrogen suppression than formestane.

As differences in plasma estrogen disposition have been reported between Japanese and Caucasian women [24], anastrozole inhibition of estrogen levels was studied in healthy, postmenopausal Japanese women. In this study plasma estradiol and estrone sulfate levels were similar in Japanese and Caucasian postmenopausal women, implying that the therapeutic benefits of anastrozole in Caucasians are predictive of the drug's effect in Japanese women.

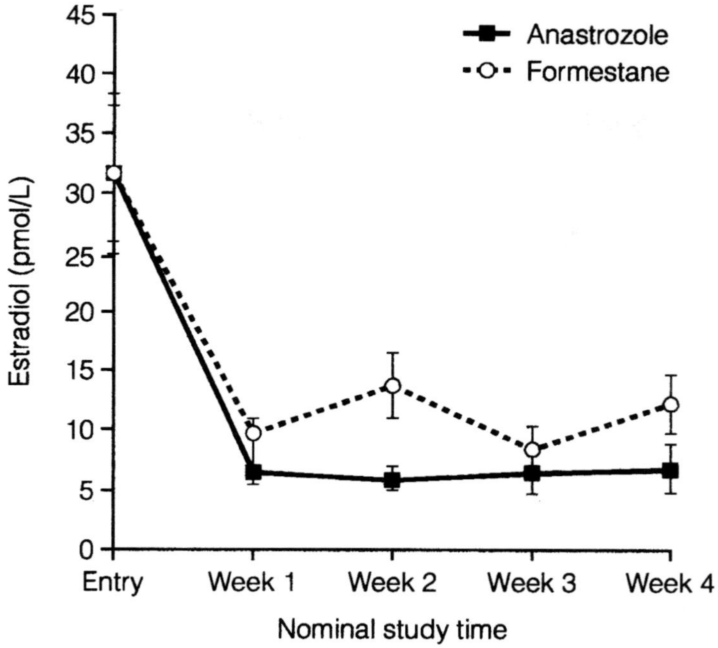


Figure 2. Mean serum estradiol concentrations of anastrozole (1 mg, by mouth, once daily) *versus* formestane (250 mg, intramuscularly, every 2 weeks). The limit of detection is 3 pM. Reprinted from [23], with permission from Oxford University Press.

*Whole-body aromatase inhibition*

Assays of peripheral aromatase activity are often used to provide a sensitive measure of the *in vivo* potency of anastrozole in postmenopausal women. In a crossover study, 12 patients progressing after tamoxifen treatment received anastrozole 1 or 10 mg once daily for 28 days [25]. *In vivo* aromatization was suppressed by 96.7 and 98.1% from baseline, respectively, and plasma levels of estradiol, estrone and estrone sulfate were reduced by ≥83.5, ≥86.5 and ≥93.5%, respectively, irrespective of dose. This study demonstrated that anastrozole was highly effective in inhibiting *in vivo* aromatization with no difference between 1 or 10 mg doses.

In another randomized crossover study, the effects of anastrozole 1 mg and letrozole 2.5 mg on total-body aromatization and plasma estrogen levels in 12 postmenopausal women with advanced breast cancer were compared [26]. A small but significantly greater decrease in the degree of suppression of estrone and estrone sulfate (but not estradiol) was observed with letrozole compared with anastrozole. However, it is becoming increasingly clear that the potency of AIs does not appear to directly correlate with efficacy [27], and small differences in estrogen suppression by these two third-generation AIs do not lead to clinically significant differences in overall efficacy when the two agents are compared directly [28].

*Intratumoral aromatase inhibition*

Despite the dramatic fall in plasma estrogen levels at the time of the menopause, postmenopausal breast tissue has the ability to maintain local concentrations of estrone and estradiol at levels that are 2–10- and 10–20-fold higher, respectively, than corresponding plasma levels [29–32]. This may be explained by uptake of estrogens from the circulation and/or *in situ* estrogen synthesis by intratumoral aromatase [33]. Such increased local estrogen levels may play a major role in breast tumor growth.

The effect of anastrozole on intratumoral aromatase has been studied by measuring breast tissue estrogen concentrations in postmenopausal women with ER-positive locally advanced breast cancer before and after 15 weeks of preoperative anastrozole therapy (1 mg, once daily) [34]. Treatment with anastrozole suppressed tissue estradiol, estrone and estrone sulfate levels by 89.0, 83.4 and 72.9%, respectively, and circulating levels by 86.1, 83.9 and 94.2%, respectively. These findings confirmed that the profound suppression of plasma estrogen levels and inhibition of total body aromatization by anastrozole (administered preoperatively) was accompanied by a similar reduction in tumor estrogen levels.

*Enzyme selectivity: interactions with other CYP enzymes and the potential for drug–drug interactions*

The effects of once-daily oral doses of anastrozole on basal and adrenocorticotrophic hormone (ACTH)-stimulated cortisol and aldosterone levels were evaluated in a 5 and 10 mg multiple-dosing study involving 19 postmenopausal women with advanced breast cancer. Following 14 days of daily dosing with anastrozole 5 or 10 mg, no significant changes in basal and ACTH-stimulated cortisol and aldosterone concentrations were observed, indicating that daily doses of anastrozole up to 10 mg have no effect on glucocorticoid or mineralocorticoid secretion [15].

Although data from preclinical studies showed that anastrozole had little or no effect on CYP-mediated metabolism, these investigations did not take into account any intracellular binding or accumulation of anastrozole in the liver. Clinical studies have shown that anastrozole has no clinically significant interactions with the CYP substrate antipyrine, cimetidine (a marker for CYP3A4) or warfarin (a marker for CYP3A4 and CYP1A2 activity) [35, 36]. Co-administration of anastrozole 1 mg with other drugs is therefore unlikely to result in drug–drug interactions mediated via CYP metabolism.

In a double-blind, placebo-controlled trial, anastrozole did not affect the pharmacokinetics of tamoxifen when the two drugs were given in combination to postmenopausal women with early breast cancer [37]. However, when anastrozole and tamoxifen were administrated concomitantly the plasma anastrozole level was lowered by 27% compared with anastrozole alone ($P < 0.001$) [38]. Although tamoxifen reduced the steady-state trough plasma concentrations of anastrozole, no significant effects on the estradiol-suppressive properties of anastrozole were observed. It was therefore concluded that the observed reduction in anastrozole levels by tamoxifen is unlikely to be of clinical significance when anastrozole and tamoxifen are administered together.

## Pharmacokinetics

Pharmacokinetic studies have shown that anastrozole 1 mg is rapidly absorbed after oral administration, with peak plasma concentrations reached after approximately 2 h [15, 21, 22]. The estimated half-life of anastrozole is approximately 40–50 h, in accordance with a once-daily dosing regimen [22]. In multiple-dosing studies, plasma concentrations of anastrozole approached steady-state levels at about 7 days of daily doses, consistent with the approximate 2-day terminal elimination half-life for anastrozole [15, 21, 22]. In these studies, steady-state levels were around 3–4-fold higher than those observed after a single dose of anastrozole.

## Anastrozole in advanced breast cancer

### *Efficacy and tolerability as a second-line agent in the treatment of advanced breast cancer*

Anastrozole was the first of the third-generation AIs to report efficacy and tolerability data from large randomized phase 3 trials in the advanced setting, as a second-line agent. The effectiveness of two oral doses of anastrozole (1 and 10 mg once daily) were compared with megestrol acetate (40 mg, four times daily) in two large multicenter trials involving postmenopausal women with advanced breast cancer who had progressed on tamoxifen [39, 40]. The designs of each trial were essentially identical (one conducted in Europe ($n = 378$) [40] and the other in North America ($n = 386$) [39]). A prospectively planned analysis of the combined results revealed that, after a median follow-up of 6 months, anastrozole (1 and 10 mg) was at least as effective as megestrol acetate for time to progression (TTP) and objective response (complete response plus partial response) [41].

Data from the mature survival analysis of the combined European and North American studies (median follow-up of 31 months) showed that at the clinical dose of 1 mg daily, anastrozole demonstrated a statistically significant survival advantage over megestrol acetate (hazard ratio, 0.78; 97.5% confidence interval, 0.6–1.0; $P < 0.025$; Fig. 3) [42]. Compared with megestrol acetate, patients treated with anastrozole 1 mg had a longer median time to death, with

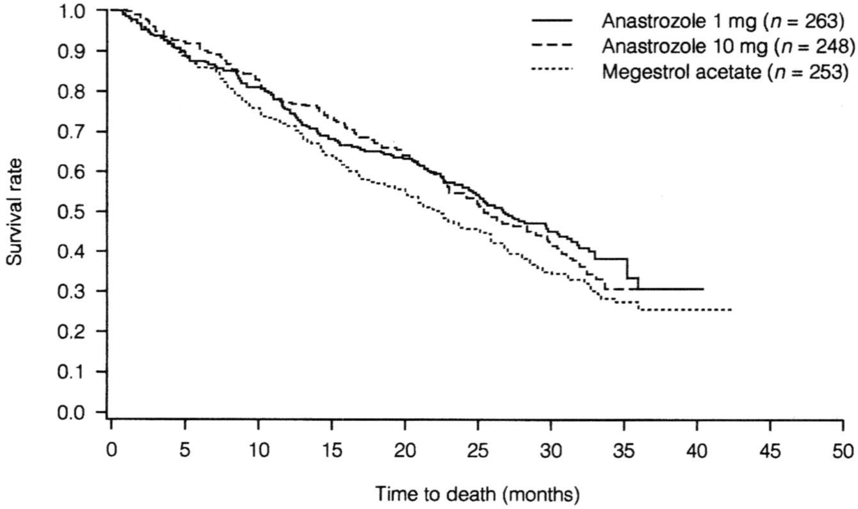


Figure 3. Kaplan–Meier survival curves for patients given anastrozole 1 and 10 mg and megestrol acetate (combined analysis of North American and European studies). Reproduced from [42]. Copyright ©1998 American Cancer Society. Reproduced with permission from Wiley-Liss, Inc., a subsidiary of John Wiley and Son, Inc.

more patients surviving for longer than 2 years (Tab. 1). Patients treated with anastrozole 10 mg also had a survival benefit over the megestrol acetate group (hazard ratio, 0.83), but this did not reach statistical significance ($P = 0.09$). Anastrozole (1 or 10 mg) was at least as effective as megestrol acetate in terms of TTP and clinical benefit (complete response+partial response+stable dis-

Table 1. Efficacy of anastrozole compared with megestrol acetate as second-line therapy: combined analysis of European and North American phase 3 trials [42]

| | Anastrozole 1 mg/day ($n = 263$) | Anastrozole 10 mg/day ($n = 248$) | Megestrol acetate 4 × 40 mg/day ($n = 253$) |
|---|---|---|---|
| Median follow-up (months) | 31 | 31 | 31 |
| Median TTP (months) | 4.8 | 5.3 | 4.6 |
| Objective response rate (%) | 12.5 | 12.5 | 12.2 |
| Clinical benefit (%) | 42.2 | 39.9 | 40.3 |
| 2-year survival (%) | 56.1 | 54.6 | 46.3 |
| Median overall survival (months) | 26.7* | 25.5 | 22.5 |

* $P < 0.025$ *versus* megestrol acetate.
CB = clinical benefit; CR = complete response; ORR = objective response rate; OS = overall survival; PR = partial survival; SD = stable disease; TTP = time to progression
Clinical benefit means complete response+partial response+stable disease ≥24 weeks; objective response rate is complete response+partial response.

ease ≥24 weeks; Tab. 1). In general, all three treatments were well tolerated in these trials, although megestrol acetate was associated with a significantly higher incidence of weight gain which continued over time.

In an additional retrospective analysis of the combined European and North American trials, a within-group comparison of patients with ($n = 237$) and without ($n = 279$) visceral metastases showed clinical benefit rates were similar between treatments [43]. objective response rates for patients in the anastrozole and megestrol acetate groups were 51.8% (72/139) *versus* 47.1% (66/140) in patients with no visceral metastases and 31.4% (39/124) *versus* 31.9% (36/113) in all patients with visceral metastases. These data show that in postmenopausal patients with advanced breast cancer and visceral metastases – who are often regarded as less likely to respond to endocrine therapy than patients without visceral metastases – anastrozole was effective as second-line therapy in advanced breast cancer.

A recent open-label trial has compared letrozole and anastrozole as second-line therapy in postmenopausal women with advanced breast cancer [28]. At a median follow-up of 5.7 months anastrozole was similar to letrozole for the primary efficacy endpoint TTP ($P = 0.92$), for time to treatment failure ($P = 0.761$), median overall survival ($P = 0.624$) and clinical benefit ($P = 0.216$). The only difference in efficacy between treatments was for the secondary endpoint, objective response, which was higher in the letrozole group compared with the anastrozole group (19.1% *versus* 12.3%, respectively; $P = 0.013$). However, when patients with confirmed hormone receptor-positive tumors only were evaluated, the two treatment groups had similar objective response rates (letrozole 17.3% *versus* anastrozole 16.8%). Taken together, these efficacy data suggest that anastrozole and letrozole are not associated with clinically relevant differences in the treatment of hormone-sensitive advanced breast cancer.

### *Efficacy and tolerability as a first-line agent in the treatment of advanced breast cancer*

Based on the utility of AIs as second-line therapy, two randomized double-blind trials were conducted to assess the effectiveness of anastrozole compared with tamoxifen, for the first-line treatment of hormone-sensitive, advanced breast cancer in postmenopausal women. Anastrozole was the first third-generation AI to be studied in this setting in two trials that were similar in design, one conducted in the United States and Canada (the so-called North American trial; anastrozole, $n = 171$; tamoxifen, $n = 182$ [44]) and the second in Europe, South America and Australia (the Tamoxifen or Arimidex Randomised Group Efficacy and Tolerability (TARGET) trial; anastrozole, $n = 340$; tamoxifen, $n = 328$ [45]). Data from the North American trial suggested that anastrozole is superior to tamoxifen as a first-line treatment of advanced breast cancer in postmenopausal women [44]. Anastrozole significantly increased TTP compared with tamoxifen (median values of 11.1 *versus* 5.6 months respectively; $P = 0.005$); the objective response rate was 21 *versus* 17%, respectively, and clinical benefit rates were 59 *versus* 46%, respectively ($P = 0.0098$) [44]. The TARGET trial further confirmed that anastrozole was at least as effective as tamoxifen in this setting with median TTP values of 8.2 and 8.3 months for anastrozole and tamoxifen, respectively ($P = 0.941$); the objective response rate was 32.9 *versus* 32.6%, respectively ($P = 0.787$), and clinical benefit rates were 56.2 and 55.5%, respectively [45].

In a prospectively planned combined analysis of these trials anastrozole was equivalent to tamoxifen in terms of TTP, objective response, clinical benefit, time to treatment failure and overall survival [46, 47] (Tab. 2). However, when the clinically relevant population was considered in a retrospective subgroup analysis of patients with ER- and/or progesterone receptor (PgR)-positive tumors, anastrozole was significantly superior to tamoxifen with respect to TTP (median values of 10.7 *versus* 6.4 months for anastrozole and tamoxifen, respectively; $P = 0.022$; Fig. 4 [46]). These analyses confirmed that receptor status was a key factor affecting the relative efficacy of anastrozole in relation to tamoxifen. An update of the safety data of the North American and TARGET trials (median duration of treatment 10.9 months for anastrozole and 8.3 months for tamoxifen) showed that anastrozole was well tolerated compared with tamoxifen, with fewer reports of vaginal bleeding and thromboembolic events in the anastrozole group compared with the tamoxifen group (Tab. 3) [47].

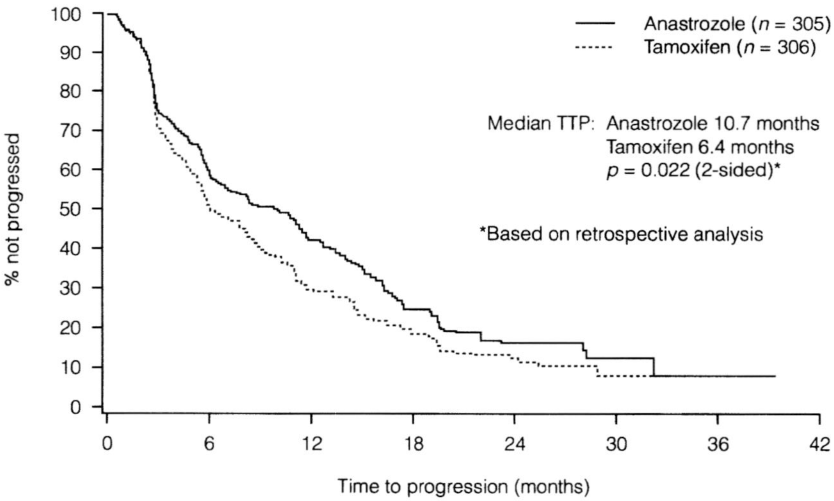


Figure 4. Kaplan–Meier curve of TTP in patients with hormone-responsive tumors receiving anastrozole 1 mg or tamoxifen (combined analysis of North American and TARGET studies) [46]. The statistical test shown was based on retrospective analysis. Reproduced from [46]. Copyright ©2001 American Cancer Society. Reproduced with permission from Wiley-Liss, Inc., a subsidiary of John Wiley and Son, Inc.

Table 2. Summary of key published efficacy results from phase 3 trials of anastrozole *versus* tamoxifen in first-line therapy of advanced breast cancer

| Study arm | Combined analysis of TARGET[a] and North American[b] trials [46, 47] | | | Single-center study [48] | | |
|---|---|---|---|---|---|---|
| | Anastrozole 1 mg ($n = 511$) | Tamoxifen ($n = 510$) | *P* value | Anastrozole 1 mg ($n = 121$) | Tamoxifen ($n = 117$) | *P* value |
| Patients with HR+ tumors (%) | 60 | | 100 | | | |
| Median follow-up (months) | 18 | | 13 | | | |
| Objective response rate (%) | 29.0 | 27.1 | NS | 36 | 26 | 0.17 |
| Clinical benefit (%) | 57.1 | 52.0 | 0.11 | 83 | 56 | 0.001 |
| Median TTP (months) | 8.5 | 7.0 | 0.10 | 18.0 | 7.0 | 0.01 |
| Median overall survival (months) | 39.2[a] | 40.1[a] | NS | 17.4 | 16.0 | 0.003 |

[a]Data reported after an extended median follow-up of 43.7 months.
CR = complete response; HR+ = hormone receptor-positive; NS = not significant; ORR = objective response rate; OS = overall survival; PR = partial survival; SD = stable disease; TARGET = Tamoxifen or Arimidex Randomised Group Efficacy and Tolerability; TTP = time to progression
Clinical benefit means complete response+partial response+stable disease ≥24 weeks; objective response rate is complete response+partial response.

Table 3. Predefined adverse events from the combined analysis of the North American and TARGET trials comparing anastrozole with tamoxifen. Reprinted from [47] with permission from Elsevier

| Adverse event | Number of events (%) | |
|---|---|---|
| | Anastrozole 1 mg/day ($n$ = 506) | Tamoxifen 20 mg/day ($n$ = 511) |
| Depression | 30 (5.9) | 36 (7.0) |
| Tumor flare | 15 (3.0) | 18 (3.5) |
| Thromboembolic disease | 27 (5.3) | 46 (9.0) |
| Gastrointestinal disturbance | 184 (36.4) | 207 (40.5) |
| Hot flashes | 139 (27.5) | 123 (24.1) |
| Vaginal dryness | 16 (3.2) | 11 (2.2) |
| Lethargy | 6 (1.2) | 17 (3.3) |
| Vaginal bleeding | 6 (1.0) | 13 (2.5) |
| Weight gain | 12 (2.4) | 8 (1.6) |

An initial survival analysis at a median of 43.7 months follow-up showed that anastrozole was not inferior to tamoxifen in terms of overall survival in both the overall population and the ER- and/or PgR-positive subgroup [47]. Although there was no improvement in survival, the favorable profile of anastrozole with respect to TTP and tolerability supports the use of anastrozole as a first-line therapy in postmenopausal women with advanced breast cancer.

An additional retrospective analysis of the combined analysis of the North American and TARGET trials showed that for patients without visceral metastases ($n$ = 528) the clinical benefit rate was 62.3% (200/321) *versus* 55.9% (166/297) for anastrozole and tamoxifen, respectively [43]. For patients with visceral metastases ($n$ = 397) the clinical benefit rates for patients in the anastrozole and tamoxifen groups were 49.5% (92/186) *versus* 46.9% (99/211). These data show that anastrozole is an effective first-line therapy in postmenopausal women with advanced breast cancer and visceral metastases.

A single-center trial has compared anastrozole with tamoxifen as first-line therapy in 238 postmenopausal patients with advanced breast cancer, all with ER-positive tumors [48]. At a median follow-up of 13.3 months, anastrozole showed significant advantages over tamoxifen for clinical benefit and overall survival (Tab. 2). Thus these data indicate that anastrozole has a survival advantage over tamoxifen in a group of patients with ER-positive tumors, a population that is likely to show the most benefit from endocrine therapy.

## *Sequencing of endocrine agents*

A further retrospective combined analysis of the North American and TARGET trials ($n$ = 1021) showed that in patients with hormone receptor-positive tumors, sequential administration of first-line anastrozole followed by tamoxifen provided clinical benefit to 48.7% of patients, while 10.1% experienced an objective response [49]. These data indicate that tumors responding to anastrozole as a first-line therapy may subsequently respond to tamoxifen as a second-line therapy. A further double-blind, crossover, substudy of the TARGET trial (the Swiss Group for Clinical Cancer Research (SAKK) 21/95 sub-trial), investigated the clinical impact of anastrozole followed by tamoxifen, compared with tamoxifen followed by anastrozole, after progression on the first treatment [50]. The results showed that overall survival from randomization for the anastrozole–tamoxifen sequence was longer than for the tamoxifen–anastrozole sequence (69.7 *versus* 59.3 months, respectively; $P = 0.1$), supporting tamoxifen as a second-line therapy after anastrozole in postmenopausal women with hormone-responsive advanced breast cancer.

## Anastrozole as adjuvant therapy for early breast cancer

### *The 'Arimidex', Tamoxifen, Alone or in Combination (ATAC) trial*

The ongoing ATAC trial is the first adjuvant breast cancer study to provide data on a third-generation AI *versus* tamoxifen in this setting. The ATAC trial has directly compared anastrozole with tamoxifen as initial adjuvant therapy in postmenopausal women with early breast cancer. A total of 9366 postmenopausal women with early disease were enrolled in this prospective, double-blind trial and were randomized to receive daily doses of anastrozole alone (1 mg), or tamoxifen alone (20 mg) or the combination. Initial and updated analyses of the ATAC trial at 33 and 47 months median follow-up showed that anastrozole significantly prolonged disease-free survival (DFS) and time to recurrence (TTR), and reduced the incidence of contralateral breast cancer (CLBC), compared with tamoxifen [51, 52]. Furthermore, anastrozole demonstrated several safety and tolerability advantages compared with tamoxifen, including a reduction in thromboembolism, ischemic cerebrovascular events and endometrial cancer. The combination arm was discontinued following the initial analysis, since it demonstrated no benefit compared with tamoxifen alone in terms of either efficacy or tolerability.

The ATAC trial completed treatment analysis, performed at a median follow-up of 68 months, further confirmed the superiority of anastrozole over tamoxifen with regards to DFS, both in the overall population and in the hormone receptor-positive subgroup (84% of the total population; Tab. 4) [53, 54]. The absolute difference in DFS between anastrozole and tamoxifen continued to increase over time in both the overall (1.5% at 3 years, 2.0% at 4 years, 2.4% at 5 years and 2.9% at 6 years; Fig. 5) and the hormone receptor-positive populations (1.6% at 3 years, 2.6% at 4 years, 2.5% at 5 years and 3.3% at 6 years), and extended beyond the completion of therapy [54].

Table 4. Major efficacy endpoints after 5 years of adjuvant treatment for early breast cancer in the ATAC trial (median follow-up of 68 months) for anastrozole compared with tamoxifen [53]

| Endpoint | All patients (anastrozole, *n* = 3125; tamoxifen, *n* = 3116) | | Hormone receptor-positive population (anastrozole, *n* = 2618; tamoxifen, *n* = 2598) | |
|---|---|---|---|---|
| | Hazard ratio (95% CI) | *P* value | Hazard ratio (95% CI) | *P* value |
| Recurrence or death | 0.87 (0.78–0.97) | 0.01 | 0.83 (0.73–0.94) | 0.005 |
| Recurrence | 0.79 (0.70–0.90) | 0.0005 | 0.74 (0.64–0.87) | 0.0002 |
| Distant recurrence | 0.86 (0.74–0.99) | 0.04 | 0.84 (0.70–1.00) | 0.06 |
| Contralateral breast cancer | 0.58 (0.38–0.88) | 0.01 | 0.47 (0.29–0.75) | 0.001 |
| All deaths | 0.97 (0.85–1.12) | 0.7 | 0.97 (0.83–1.14) | 0.7 |
| Breast cancer deaths | 0.88 (0.74–1.05) | 0.2 | 0.87 (0.70–1.09) | 0.2 |
| Non-breast cancer deaths | 1.13 (0.91–1.40) | 0.3 | 1.10 (0.87–1.40) | 0.4 |

Contralateral breast cancer includes ductal carcinoma *in situ*.

TTR was also significantly prolonged in patients receiving anastrozole compared with those treated with tamoxifen, both in the overall population and hormone receptor-positive subgroup (Tab. 4) [53, 54]. The difference in TTR between anastrozole and tamoxifen treatment arms continued to increase over time, so that at year 6 the absolute difference was 3.4% for the overall popu-

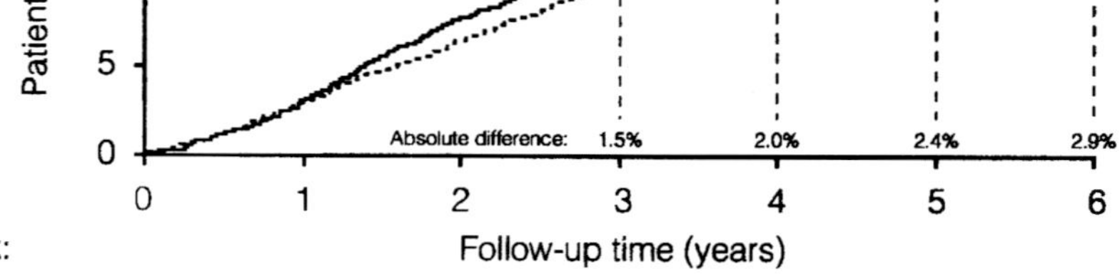


| At risk: | 0 | 1 | 2 | 3 | 4 | 5 | 6 |
|---|---|---|---|---|---|---|---|
| Anastrozole | 3125 | 3004 | 2874 | 2757 | 2645 | 2350 | 984 |
| Tamoxifen | 3116 | 2992 | 2835 | 2709 | 2575 | 2273 | 933 |

Figure 5. DFS in the overall (intent-to-treat) population at 1–6 years of treatment in the ATAC trial, showing the absolute differences for years 3–6. CI, confidence interval. Data from A. Howell, unpublished observations.

lation and 3.7% for patients with hormone receptor-positive tumors. In addition, for both the overall and receptor-positive population the completed treatment analysis confirmed the benefits of anastrozole over tamoxifen with regards to a significant reduction in the incidence of CLBC and also showed a significant reduction in invasive CLBC, compared with tamoxifen alone.

For the first time, the completed treatment analysis demonstrated that the DFS and TTR benefits demonstrated by anastrozole over tamoxifen resulted in a benefit in time to distant recurrence (Tab. 4) [53]. After 3 years, an absolute difference emerged that continued to increase over time in the overall (0.7% at 3 years, 1.3% at 4 years, 1.5% at 5 years and 2.1% at 6 years) and hormone receptor-positive (0.7% at 3 years, 1.3% at 4 years, 1.3% at 5 years, 1.9% at 6 years) populations [54]. The significant reduction in recurrence and distant recurrence demonstrated by anastrozole strongly suggests that an eventual benefit in breast cancer survival will be observed. In the survival analysis, overall survival was similar for both anastrozole and tamoxifen (Tab. 4) [53], demonstrating that anastrozole maintains the established survival benefit observed for tamoxifen; however, there was a 12% lower breast cancer death rate with anastrozole, but this did not reach statistical significance. A similar trend was observed for the hormone receptor-positive population. As the trial population had a relatively good prognosis (61% of patients were known to be lymph node-negative and 64% had tumors of ≤2 cm), it is too early to expect a survival difference. Other large adjuvant trials have taken up to 7 years before a survival benefit could be established for tamoxifen *versus* placebo.

As ≥90% of patients had completed treatment, the safety and tolerability analysis at 5 years can be considered final. Anastrozole was significantly better tolerated than tamoxifen with respect to endometrial cancer, vaginal bleeding and discharge, hot flashes, ischemic cerebrovascular events, venous thromboembolic events and deep venous thromboembolic events (Tab. 5) [53]. Indeed, compared with anastrozole, women treated with tamoxifen had a substantially higher hysterectomy rate (1.3 *versus* 5.1% for anastrozole and tamoxifen, respectively) or hysterectomy for malignancy (0.3 *versus* 0.9%, respectively) [55]. Although fewer fractures of the spine and at sites other than the hip, spine and wrist/colles were reported in the tamoxifen group compared with the anastrozole group, fractures at the hip and wrist/colles were similar between the two treatment groups. Furthermore, the relative risk of fracture was shown to stabilize after 24 months with no subsequent increase over time [56]. The ATAC completed treatment analysis showed that withdrawals due to adverse events were significantly less common with anastrozole (11.1% (344/3092)) than with tamoxifen (14.3% (442/3094); $P = 0.0002$) and that drug-related serious adverse events were also significantly less common with anastrozole (4.7% (146/3092)) than with tamoxifen (9.0% (271/3094); $P < 0.0001$). Consequently, overall the risk/benefit profile remains in favor of anastrozole. The higher rates of recurrence, adverse events and treatment withdrawals associated with tamoxifen, and the substantial benefit of anastrozole, support the approach of offering anastrozole at the earliest opportunity in the adjuvant setting.

Table 5. Predefined adverse events on treatment or within 14 days of discontinuation after 5 years of adjuvant treatment for early breast cancer in the ATAC trial for anastrozole compared with tamoxifen. CI, confidence interval. Reprinted from [53] with permission from Elsevier

| Adverse event | Number of patients (%) | | Odds ratio of anastrozole *versus* tamoxifen (95% CI) | *P* value |
|---|---|---|---|---|
| | Anastrozole ($n = 3092$) | Tamoxifen ($n = 3094$) | | |
| Hot flashes | 1104 (35.7) | 1264 (40.9) | 0.80 (0.73–0.89) | <0.0001 |
| Nausea and vomiting | 393 (12.7) | 384 (12.4) | 1.03 (0.88–1.19) | 0.7 |
| Fatigue/tiredness | 575 (18.6) | 544 (17.6) | 1.07 (0.94–1.22) | 0.3 |
| Mood disturbances | 597 (19.3) | 554 (17.9) | 1.10 (0.97–1.25) | 0.2 |
| Arthralgia | 1100 (35.6) | 911 (29.4) | 1.32 (1.19–1.47) | <0.0001[c] |
| Vaginal bleeding | 167 (5.4) | 317 (10.2) | 0.50 (0.41–0.61) | <0.0001 |
| Vaginal discharge | 109 (3.5) | 408 (13.2) | 0.24 (0.19–0.30) | <0.0001 |
| Endometrial cancer[a] | 5 (0.2) | 17 (0.8) | 0.29 (0.11–0.80) | 0.02 |
| Fractures[b] | 340 (11) | 237 (7.7) | 1.49 (1.25–1.77) | <0.0001[c] |
| Hip | 37 (1.2) | 31 (1.0) | 1.20 (0.74–1.93) | 0.5 |
| Spine | 45 (1.5) | 27 (0.9) | 1.68 (1.04–2.71) | 0.03[c] |
| Wrist/colles | 72 (2.3) | 63 (2.0) | 1.15 (0.81–1.64) | 0.4 |
| All other sites[d] | 220 (7.1) | 142 (4.6) | 1.59 (1.28–1.98) | <0.0001[c] |
| Ischemic cardiovascular disease | 127 (4.1) | 104 (3.4) | 1.23 (0.95–1.60) | 0.1 |
| Ischemic cerebrovascular events | 62 (2.0) | 88 (2.8) | 0.70 (0.50–0.97) | 0.03 |
| Venous thromboembolic events | 87 (2.8) | 140 (4.5) | 0.61 (0.47–0.80) | 0.0004 |
| Deep-venous thrombo-embolic events | 48 (1.6) | 74 (2.4) | 0.64 (0.45–0.93) | 0.02 |
| Cataracts | 182 (5.9) | 213 (6.9) | 0.85 (0.69–1.04) | 0.1 |

[a] $n = 2229$, 2236 for anastrozole and tamoxifen, respectively (excluding patients with hysterectomy at baseline), recorded at any time.
[b] Patients with one or more fractures occurring at any time before recurrence (includes patients no longer receiving treatment).
[c] In favor of tamoxifen.
[d] Patients may have had one or more fracture at different sites.

## *Switching studies with anastrozole*

For patients who are currently receiving tamoxifen, emerging results from the Arimidex-Nolvadex (ARNO 95) phase 3 trials involving the German Adjuvant Breast Group (GABG) and the Austrian Breast and Colorectal Cancer Study Group (ABCSG 8) have demonstrated the benefits of switching to anastrozole after prior tamoxifen treatment. Following 2 years of adjuvant tamoxifen, postmenopausal patients with early breast cancer were randomized to receive a further 3 years of tamoxifen (20 or 30 mg/day; $n = 1606$) or to switch to 3 years of anastrozole (1 mg/day; $n = 1618$), for a total duration of 5 years of endocrine therapy. At a median follow-up of 28 months there was a hazard ratio of 0.6 in favor of anastrozole for the occurrence of an event (95% confidence interval, 0.44–0.81; $P = 0.0009$), representing a risk reduction of 40% for patients receiving anastrozole [57]. These data are in agreement with the findings of a smaller study by the Italian Tamoxifen Anastrozole (ITA) group which found a significant difference in event-free survival ($P = 0.0002$) and recurrence-free survival ($P = 0.001$) between treatment groups, which favored switching to anastrozole [58]. Overall, switching to anastrozole was generally better tolerated than tamoxifen.

## Anastrozole as preoperative endocrine therapy for breast cancer

Successful preoperative systemic therapy aims to reduce tumor mass and downstage the tumor, thereby allowing breast-conserving surgery in patients with large operable breast cancer and rendering inoperable tumors resectable. As discussed previously, anastrozole inhibits total-body aromatization and also reduces intratumoral estrogen levels when administered preoperatively [34], suggesting a potential role in the preoperative setting. In a small randomized, double-blind, single-center study preoperative anastrozole was efficacious in reducing tumor volume in postmenopausal women with newly diagnosed, ER-rich, locally advanced or large operable breast tumors [59]. Treatment with anastrozole (1 or 10 mg) resulted in 15/17 patients initially requiring a mastectomy becoming eligible for breast-conserving surgery. In addition, anastrozole renders locally advanced breast tumors operable through mastectomy, regardless of tumor erbB2 or Ki67 status [60]. Further studies in postmenopausal patients with locally advanced breast cancer have confirmed the benefit of preoperative anastrozole (1 mg/day for 3 months) over tamoxifen in terms of tumor shrinkage [61, 62] and objective response [63].

The Immediate Preoperative ‘Arimidex’, Tamoxifen or Combined with Tamoxifen (IMPACT) trial compared 3 months of anastrozole with tamoxifen, or the combination, as preoperative treatment of ER-positive operable breast cancer (including locally advanced tumors) in postmenopausal women ($n = 330$) [64]. In the overall population, anastrozole and tamoxifen were equally effective, with objective response achieved in 37.2 and 36.1% of patients, respectively. Of the 124 patients requiring a mastectomy at baseline (the clinically relevant target population for preoperative treatment), objective response was achieved in 39.1 and 27.8% of patients in the anastrozole and tamoxifen groups, respectively. Treatment with preoperative anastrozole was, therefore, more likely to downstage tumors, enabling more patients in the anastrozole group to have breast-conserving surgery compared with tamoxifen alone (45.7 *versus* 22.2%, respectively).

The efficacy of preoperative anastrozole therapy has been confirmed in the Preoperative Arimidex Compared with Tamoxifen (PROACT) trial [65]. This trial evaluated the efficacy of anastrozole ($n = 228$) compared with tamoxifen ($n = 223$) as preoperative therapy in postmenopausal women with large, operable or potentially operable, locally advanced, hormone receptor-positive breast tumors. At baseline, 14.2% of patients had tumors assessed as suitable for breast-conserving surgery, 78.3% for mastectomy and 7.3% had inoperable tumors. Significant improvements in actual surgery were observed in 262 patients who were treated with anastrozole therapy alone (no additional preoperative chemotherapy) compared with tamoxifen. For patients requiring mastectomy at study entry objective response assessed by ultrasound was 36.6 and 24.2% for anastrozole and tamoxifen, respectively ($P = 0.03$). Treatment with preoperative anastrozole was therefore more likely to downstage tumors than tamoxifen, which was reflected in significantly more of the anastrozole-treated patients, demonstrating a reduction in the extent of surgery required compared with tamoxifen (43 *versus* 31%, respectively; $P = 0.04$).

In a prospectively planned combined analysis of results of the IMPACT and PROACT trials ($n = 535$), significantly more patients treated with preoperative anastrozole compared with tamoxifen experienced improvement in both feasible (47 *versus* 35%; $P = 0.021$) and actual (43 *versus* 31%; $P = 0.019$) surgery [66]. Preoperative anastrozole therapy has also been shown to be well tolerated [62, 64, 65]. Together these data support the role of anastrozole as an effective preoperative therapy for postmenopausal patients with hormone receptor-positive, large or locally advanced breast tumors.

## Anastrozole for chemoprevention

Estrogens have been implicated in both the initiation and prevention of breast cancer. Although several prevention trials have shown that tamoxifen can reduce the incidence of breast cancer in high-risk women [67, 68], it is associated with an increased risk of thromboembolic disease and endometrial cancer [69]. The superiority of anastrozole over tamoxifen in terms of both efficacy and toxicity in advanced disease as well as in the preoperative and adjuvant setting has led to the launch of the International Breast Cancer Intervention Study (IBIS) II trial [70]. Since tamoxifen shows a 50% reduction in the occurrence of tumors in hormone-receptor-positive patients compared with placebo [71], the findings from the ATAC study suggest that anastrozole treatment might prevent 70–80% of hormone receptor-positive tumors in women at high risk of breast cancer. The IBIS II trial [70] will compare anastrozole with placebo in 6000 high-risk postmenopausal women who are not receiving hormone-replacement therapy and who are at increased risk of developing breast cancer. The second stratum of the trial will compare anastrozole *versus* tamoxifen in 4000 women with locally excised ductal carcinoma *in situ*.

## Long-term safety and tolerability of AIs

The studies described have confirmed that anastrozole has fewer thromboembolism and ischemic cerebrovascular events compared with tamoxifen, and does not demonstrate androgenic, progestogenic or estrogenic effects such as weight gain, acne or hypertrichosis. Although estrogen deprivation has the potential to alter lipid profiles detrimentally, additional studies have shown that this is not the case with anastrozole [72–75]. Whereas switching from adjuvant tamoxifen to anastrozole was associated with a higher incidence of lipid disorders in the ITA trial, this may be due to the effects of discontinuing tamoxifen, which is known to have a beneficial effect on the lipid profile [58]. Although anastrozole was associated with a higher incidence of joint symptoms and fractures compared with tamoxifen in the ATAC trial [53], risk ratios remained constant over the treatment period. Indeed, an analysis of 6-monthly fracture rates over time (between 6–48 months of treatment) showed that, with anastrozole, fracture rates stabilized after an initial increase during the first 2 years and the relative risk *versus* tamoxifen did not worsen with continued treatment [56]. Indirect comparisons of fracture rates between the ATAC trial and other major trials [67, 68, 76–79] show that fracture rates in the ATAC trial fall within the broad range of those reported in other large trials or surveys, suggesting that any increase in fracture rates associated with anastrozole is modest. Bone loss associated with AI treatment is now recognized as a preventable and treatable condition and adjuvant bisphosphonates may become a more standard component of the treatment of women with early-stage breast cancer in the future [80]. Overall, data obtained in the clinical studies described in this chapter show that anastrozole is well tolerated and has an improved tolerability profile over tamoxifen.

## Conclusions

Preclinical studies demonstrate that anastrozole is a potent and highly selective AI with a pharmacological profile suitable for the treatment of breast cancer. Anastrozole was the first of the third-generation AIs to report efficacy and tolerability data from large randomized phase 3 trials involving patients with advanced disease (as a second- or first-line agent), or as an adjuvant treatment for early breast cancer. In clinical trials, anastrozole has been shown to be superior to megestrol acetate, in terms of survival and adverse effects, as a second-line therapy in postmenopausal women with ER-positive and/or PgR-positive advanced breast cancer. Phase 3 clinical trials have also demonstrated that anastrozole significantly prolongs the time to tumor progression compared with tamoxifen as a first-line therapy for ER- and/or PgR-positive advanced breast cancer in postmenopausal women. Therefore, as well as being established as a second-line treatment for advanced breast cancer, the improved risk/benefit profile of anastrozole over tamoxifen means that anastrozole is now also recognized as an alternative to tamoxifen in the first-line setting.

In the adjuvant setting, results of the ATAC trial have shown that anastrozole is superior to tamoxifen in terms of DFS, TTR, distant time to recurrence and prevention of CLBC in postmenopausal women with early ER-positive breast cancer. Findings also indicate that, overall, anastrozole has a more favorable tolerability profile than tamoxifen. Although longer follow-up may be required to assess fully the long-term effects of anastrozole on bone mineral density and overall survival, overall results thus far are extremely promising and may be just as significant as the findings first seen with tamoxifen over 20 years ago. Since there are differences in the pharmacological profiles of the third-generation AIs and it is unknown if the AIs are interchangeable in clinical practice, data from the ATAC trial may only be applicable to anastrozole. Results from the ATAC trial suggest, therefore, that anastrozole should be considered as the preferred initial adjuvant therapy for the treatment in postmenopausal women with hormone receptor-positive breast cancer during the first 5 years following surgery, when most breast cancer recurrences occur [71, 81]. For patients with early breast cancer who are currently receiving tamoxifen, trials evaluating anastrozole after 2–3 years of adjuvant tamoxifen compared with continuing tamoxifen have demonstrated the benefits of switching to anastrozole. Consequently, in the adjuvant setting, anastrozole is the only AI with conclusive evidence demonstrating superior efficacy and tolerability *versus* tamoxifen in both newly diagnosed patients and in those patients already receiving adjuvant tamoxifen. Based on results from trials with adjuvant anastrozole, the American Society of Clinical Oncology Technology Assessment on the use of AIs has recommended that the optimal adjuvant therapy for postmenopausal women with hormone receptor-positive breast cancer should now include an AI as initial therapy or after treatment with tamoxifen, in order to lower the risk of tumor recurrence [80].

## References

1 Chen S (1998) Aromatase and breast cancer. *Front Biosci* 3: d922–d933
2 Wakeling AE (1990) Mechanisms of growth regulation of human breast cancer. *Baillieres Clin Endocrinol Metab* 4: 51–66
3 Simpson ER, Dowsett M (2002) Aromatase and its inhibitors: significance for breast cancer therapy. *Recent Prog Horm Res* 57: 317–338
4 Howell A, Dowsett M (1997) Recent advances in endocrine therapy of breast cancer. *BMJ* 315: 863–866
5 Gruber CJ, Tschugguel W, Schneeberger C, Huber JC (2002) Production and actions of estrogens. *N Engl J Med* 346: 340–352
6 Santen RJ, Santner S, Davis B, Veldhuis J, Samojlik E, Ruby E (1978) Aminoglutethimide inhibits extraglandular estrogen production in postmenopausal women with breast carcinoma. *J Clin Endocrinol Metab* 47: 1257–1265
7 Coombes RC (1998) Aromatase inhibitors and their use in the adjuvant setting. *Recent Results Cancer Res* 152: 277–284
8 Gale KE, Andersen JW, Tormey DC, Mansour EG, Davis TE, Horton J, Wolter JM, Smith TJ,

Cummings FJ (1994) Hormonal treatment for metastatic breast cancer. An Eastern Cooperative Oncology Group Phase III trial comparing aminoglutethimide to tamoxifen. *Cancer* 73: 354–361
9 Lipton A, Santen RJ (1974) Proceedings: Medical adrenalectomy using aminoglutethimide and dexamethasone in advanced breast cancer. *Cancer* 33: 503–512
10 Goss PE, Powles TJ, Dowsett M, Hutchison G, Brodie AM, Gazet JC, Coombes RC (1986) Treatment of advanced postmenopausal breast cancer with an aromatase inhibitor, 4-hydroxyandrostenedione: phase II report. *Cancer Res* 46: 4823–4826
11 Wiseman LR, McTavish D (1993) Formestane. A review of its pharmacodynamic and pharmacokinetic properties and therapeutic potential in the management of breast cancer and prostatic cancer. *Drugs* 45: 66–84
12 Perez-Carrion R, Alberola Candel V, Calabresi F (1996) Comparison of the selective aromatase inhibitor formestane with tamoxifen as first-line hormonal therapy in postmenopausal women with advanced breast cancer. *Ann Oncol* 5 (suppl. 7): S19–S24
13 Smith IE, Dowsett M (2003) Aromatase inhibitors in breast cancer. *N Engl J Med* 348: 2431–2442
14 Dukes M, Edwards PN, Large M, Smith IK, Boyle T (1996) The preclinical pharmacology of "Arimidex" (anastrozole; ZD1033)-a potent, selective aromatase inhibitor. *J Steroid Biochem Mol Biol* 58: 439–445
15 Plourde PV, Dyroff M, Dukes M (1994) Arimidex: a potent and selective fourth-generation aromatase inhibitor. *Breast Cancer Res Treat* 30: 103–111
16 Grimm SW, Dyroff MC (1997) Inhibition of human drug metabolizing cytochromes p450 by anastrozole, a potent and selective inhibitor of aromatase. *Drug Metab Dispos* 25: 598–602
17 Hozumi Y, Hakamata Y, Sasanuma H, Ogura S, Nagai H (2002) Effects of anastrozole on lipid metabolism compared with tamoxifen in rats. *Breast Cancer Res Treat* 76: 131–136
18 Yue W, Zhou D, Chen S, Brodie A (1994) A new nude mouse model for postmenopausal breast cancer using MCF-7 cells transfected with the human aromatase gene. *Cancer Res* 54: 5092–5095
19 Lu Q, Liu Y, Long BJ, Grigoryev D, Gimbel M, Brodie A (1999) The effect of combining aromatase inhibitors with antiestrogens on tumor growth in a nude mouse model for breast cancer. *Breast Cancer Res Treat* 57: 183–192
20 Dukes M (1997) The relevance of preclinical models to the treatment of postmenopausal breast cancer. *Oncology* 54 (suppl. 2): 6–10
21 Plourde PV, Dyroff M, Dowsett M (1995) Arimidex: a new oral, once-a-day aromatase inhibitor. *J Steroid Biochem Mol Biol* 53: 175–179
22 Yates RA, Dowsett M, Fisher GV, Selen A, Wyld PJ (1996) Arimidex (ZD1033): a selective, potent inhibitor of aromatase in postmenopausal female volunteers. *Br J Cancer* 73: 543–548
23 Vorobiof DA, Kleeberg UR, Perez-Carrion R, Dodwell DJ, Robertson JF, Calvo L, Dowsett M, Clack G (1999) A randomized, open, parallel-group trial to compare the endocrine effects of oral anastrozole (Arimidex) with intramuscular formestane in postmenopausal women with advanced breast cancer. *Ann Oncol* 10: 1219–1225
24 Dowsett M, Donaldson K, Tsuboi M, Wong J, Yates R (2000) Effects of the aromatase inhibitor anastrozole on serum oestrogens in Japanese and Caucasian women. *Cancer Chemother Pharmacol* 46: 35–39
25 Geisler J, King N, Dowsett M, Ottestad L, Lundgren S, Walton P, Kormeset PO, Lønning PE (1996) Influence of anastrozole (Arimidex), a selective, non-steroidal aromatase inhibitor, on *in vivo* aromatisation and plasma oestrogen levels in postmenopausal women with breast cancer. *Br J Cancer* 74: 1286–1291
26 Geisler J, Haynes B, Anker G, Dowsett M, Lønning PE (2002) Influence of letrozole and anastrozole on total body aromatization and plasma estrogen levels in postmenopausal breast cancer patients evaluated in a randomized, cross-over study. *J Clin Oncol* 20: 751–757
27 Sainsbury R (2004) Aromatase inhibition in the treatment of advanced breast cancer: is there a relationship between potency and clinical efficacy? *Br J Cancer* 90: 1733–1739
28 Rose C, Vtoraya O, Pluzanska A, Davidson N, Gershanovich M, Thomas R, Johnson S, Caicedo JJ, Gervasio H, Manikhas G et al. (2003) An open randomised trial of second-line endocrine therapy in advanced breast cancer. Comparison of the aromatase inhibitors letrozole and anastrozole. *Eur J Cancer* 39: 2318–2327
29 Fishman J, Nisselbaum JS, Menendez-Botet CJ, Schwartz MK (1977) Estrone and estradiol content in human breast tumors: relationship to estradiol receptors. *J Steroid Biochem* 8: 893–896
30 Pasqualini JR, Chetrite G, Blacker C, Feinstein MC, Delalonde L, Talbi M, Maloche C (1996) Concentrations of estrone, estradiol, and estrone sulfate and evaluation of sulfatase and aromatase

activities in pre- and postmenopausal breast cancer patients. *J Clin Endocrinol Metab* 81: 1460–1464

31 Vermeulen A, Deslypere JP, Paridaens R, Leclercq G, Roy F, Heuson JC (1986) Aromatase, 17 beta-hydroxysteroid dehydrogenase and intratissular sex hormone concentrations in cancerous and normal glandular breast tissue in postmenopausal women. *Eur J Cancer Clin Oncol* 22: 515–525

32 van Landeghem AA, Poortman J, Nabuurs M, Thijssen JH (1985) Endogenous concentration and subcellular distribution of androgens in normal and malignant human breast tissue. *Cancer Res* 45: 2907–2912

33 Geisler J (2003) Breast cancer tissue estrogens and their manipulation with aromatase inhibitors and inactivators. *J Steroid Biochem Mol Biol* 86: 245–253

34 Geisler J, Detre S, Berntsen H, Ottestad L, Lindtjorn B, Dowsett M, Einstein LP (2001) Influence of neoadjuvant anastrozole (Arimidex) on intratumoral estrogen levels and proliferation markers in patients with locally advanced breast cancer. *Clin Cancer Res* 7: 1230–1236

35 Dowsett M, Yates R, Wong Y and on behalf of the Arimidex Study Group (1998) 'Arimidex' (anastrozole): lack of interactions with tamoxifen, antipyrine, cimetidine and warfarin. *Eur J Cancer* 34: S39–S40

36 Yates RA, Wong J, Seiberling M, Merz M, Marz W, Nauck M (2001) The effect of anastrozole on the single-dose pharmacokinetics and anticoagulant activity of warfarin in healthy volunteers. *Br J Clin Pharmacol* 51: 429–435

37 Dowsett M, Tobias JS, Howell A, Blackman GM, Welch H, King N, Ponzone R, von Euler M, Baum M (1999) The effect of anastrozole on the pharmacokinetics of tamoxifen in postmenopausal women with early breast cancer. *Br J Cancer* 79: 311–315

38 Dowsett M, Cuzick J, Howell A, Jackson I (2001) Pharmacokinetics of anastrozole and tamoxifen alone, and in combination, during adjuvant endocrine therapy for early breast cancer in postmenopausal women: a sub-protocol of the 'Arimidex', Tamoxifen Alone or in Combination' (ATAC) trial. *Br J Cancer* 85: 317–324

39 Buzdar AU, Jones SE, Vogel CL, Wolter J, Plourde P, Webster A (1997) A phase III trial comparing anastrozole (1 and 10 milligrams), a potent and selective aromatase inhibitor, with megestrol acetate in postmenopausal women with advanced breast carcinoma. Arimidex Study Group. *Cancer* 79: 730–739

40 Jonat W, Howell A, Blomqvist C, Eiermann W, Winblad G, Tyrrell C, Mauriac L, Roche H, Lundgren S, Hellmund R et al. (1996) A randomised trial comparing two doses of the new selective aromatase inhibitor anastrozole (Arimidex) with megestrol acetate in postmenopausal patients with advanced breast cancer. *Eur J Cancer* 32A: 404–412

41 Buzdar A, Jonat W, Howell A, Jones SE, Blomqvist C, Vogel CL, Eiermann W, Wolter JM, Azab M, Webster A et al. (1996) Anastrozole, a potent and selective aromatase inhibitor, *versus* megestrol acetate in postmenopausal women with advanced breast cancer: results of overview analysis of two phase III trials. Arimidex Study Group. *J Clin Oncol* 14: 2000–2011

42 Buzdar AU, Jonat W, Howell A, Jones SE, Blomqvist CP, Vogel CL, Eiermann W, Wolter JM, Steinberg M, Webster A et al. (1998) Anastrozole *versus* megestrol acetate in the treatment of postmenopausal women with advanced breast carcinoma: results of a survival update based on a combined analysis of data from two mature phase III trials. Arimidex Study Group. *Cancer* 83: 1142–1152

43 Howell A, Robertson JF, Vergote I (2003) A review of the efficacy of anastrozole in postmenopausal women with advanced breast cancer with visceral metastases. *Breast Cancer Res Treat* 82: 215–222

44 Nabholtz JM, Buzdar A, Pollak M, Harwin W, Burton G, Mangalik A, Steinberg M, Webster A, von Euler M (2000) Anastrozole is superior to tamoxifen as first-line therapy for advanced breast cancer in postmenopausal women: results of a North American multicenter randomized trial. Arimidex Study Group. *J Clin Oncol* 18: 3758–3767

45 Bonneterre J, Thürlimann B, Robertson JF, Krzakowski M, Mauriac L, Koralewski P, Vergote I, Webster A, Steinberg M, von Euler M (2000) Anastrozole *versus* tamoxifen as first-line therapy for advanced breast cancer in 668 postmenopausal women: results of the Tamoxifen or Arimidex Randomized Group Efficacy and Tolerability study. *J Clin Oncol* 18: 3748–3757

46 Bonneterre J, Buzdar A, Nabholtz JM, Robertson JF, Thürlimann B, von Euler M, Sahmoud T, Webster A, Steinberg M, Arimidex Writing Committee et al. (2001) Anastrozole is superior to tamoxifen as first-line therapy in hormone receptor positive advanced breast carcinoma. *Cancer* 92: 2247–2258

47 Nabholtz JM, Bonneterre J, Buzdar A, Robertson JFR, Thürlimann B, for the Arimidex Writing Committee on behalf of the Investigators (2003) Anastrozole (Arimidex™) *versus* tamoxifen as first-line therapy for advanced breast cancer in postmenopausal women: survival analysis and updated safety results. *Eur J Cancer* 39: 1684–1689
48 Milla-Santos A, Milla L, Portella J, Rallo L, Pons M, Rodes E, Casanovas J, Puig-Gali M (2003) Anastrozole *versus* tamoxifen as first-line therapy in postmenopausal patients with hormone-dependent advanced breast cancer: a prospective, randomized, phase III study. *Am J Clin Oncol* 26: 317–322
49 Thürlimann B, Robertson JF, Nabholtz JM, Buzdar A, Bonneterre J, Arimidex Study Group (2003) Efficacy of tamoxifen following anastrozole ('Arimidex') compared with anastrozole following tamoxifen as first-line treatment for advanced breast cancer in postmenopausal women. *Eur J Cancer* 39: 2310–2317
50 Thürlimann B, Hess D, Köberle D, Senn I, Ballabeni P, Pagani O, Perey L, Aebi S, Rochiltz C, Goldhirsch A et al. (2004) Anastrozole ('Arimidex') *versus* tamoxifen as first-line therapy in postmenopausal women with advanced breast cancer: results of the double-blind cross-over SAKK trial 21/95 – a sub-study of the TARGET (Tamoxifen or 'Arimidex' Randomized Group Efficacy and Tolerability) trial. *Breast Cancer Res Treat* 85: 247–254
51 ATAC Trialists' Group (2002) Anastrozole alone or in combination with tamoxifen *versus* tamoxifen alone for adjuvant treatment of postmenopausal women with early breast cancer: first results of the ATAC randomised trial. *Lancet* 359: 2131–2139
52 ATAC Trialists' Group (2003) Anastrozole alone or in combination with tamoxifen *versus* tamoxifen alone for adjuvant treatment of postmenopausal women with early-stage breast cancer. Results of the ATAC (Arimidex, Tamoxifen, Alone or in Combination) trial efficacy and safety update analyses. *Cancer* 98: 1802–1810
53 ATAC Trialists' Group (2005) Results of the ATAC (Arimidex, Tamoxifen, Alone or in Combination) trial after completion of 5 years' adjuvant treatment for breast cancer. *Lancet* 365: 60–62
54 Howell A (2004) The ATAC ('Arimidex', Tamoxifen, Alone or in Combination) trial in postmenopausal women with early breast cancer – updated efficacy results based on a median follow-up of 5 years. *Breast Cancer Res Treat* 88 (suppl. 1): S7
55 Duffy S (2005) Gynecological adverse events including hysterectomy with anastrozole tamoxifen: data from the ATAC ('Arimidex', Tamoxifen, Alone or in Combination) trial. *Clin Oncol* 23: 58
56 Locker GY, Eastell R (2003) The time course of bone fractures observed in the ATAC ('Arimidex', Tamoxifen, Alone or in Combination) trial. *Proc Am Soc Clin Oncol* 22: 25
57 Jakesz R, Kaufmann M, Gnant M, Jonat W, Mittleboek M, Greil R, Tausch C, Hilfrich J, Kwasny W, Samonigg H et al. (2004) Benefits of switching postmenopausal women with hormone-sensitive early breast cancer to anastrozole after 2 years adjuvant tamoxifen: combined results from 3,123 women enrolled in the ABCSG Trial 8 and the ARNO 95 Trial. *Breast Cancer Res Treat* 88 (suppl. 1): S7
58 Boccardo F, Rubagotti A, Amoroso D, Mesiti M, Massobrio M, Benedetto C, Porpiglia M, Rinaldini M, Paladini G, Distante V et al. (2003) Anastrozole appears to be superior to tamoxifen in women already receiving adjuvant tamoxifen treatment. *Breast Cancer Res Treatment* 82 (suppl. 1): S6–S7
59 Dixon JM, Renshaw L, Bellamy C, Stuart M, Hoctin-Boes G, Miller WR (2000) The effects of neoadjuvant anastrozole (Arimidex) on tumor volume in postmenopausal women with breast cancer: a randomized, double-blind, single-center study. *Clin Cancer Res* 6: 2229–2235
60 Milla-Santos A, Milla L, Calvo N, Portella J, Rallo L, Casanovas J, Pons M (2003) Anastrozole is an effective neoadjuvant therapy for patients with hormone-dependent, locally-advanced breast cancer irrespective of cerbB2. *Proc Am Soc Clin Oncol* 22: 39
61 Anderson TJ, Dixon JM, Stuart M, Sahmoud T, Miller WR (2002) Effect of neoadjuvant treatment with anastrozole on tumour histology in postmenopausal women with large operable breast cancer. *Br J Cancer* 87: 334–338
62 Semiglazov VF, Semiglazov VV, Ivanov VG, Ziltsova EK, Dashyan GA, Kletzel A, Bozhok AA, Nurgaziev KS, Tzyrlina EV, Berstein LM (2003) Anastrozole (A) vs tamoxifen (T) vs combine (A + T) as neoadjuvant endocrine therapy of postmenopausal breast cancer patients. *Proc Am Soc Clin Oncol* 22: 880
63 Milla-Santos A, Milla L, Calvo N, Portella J, Rallo L, Casanovas JM, Pons M, Rodes J (2004)

Anastrozole as neoadjuvant therapy for patients with hormone-dependent, locally-advanced breast cancer. *Anticancer Res* 24: 1315–1318
64 Smith I, Dowsett M, on behalf of the IMPACT Trialists (2003) Comparison of anastrozole vs tamoxifen alone and in combination as neoadjuvant treatment of estrogen receptor-positive (ER+) operable breast cancer in postmenopausal women: the IMPACT trial. *Breast Cancer Res Treat* 82 (suppl. 1): S6
65 Cataliotti L, Buzdar A, Noguchi S, Bines J (2004) Efficacy of pre-operative Arimidex (anastrozole) compared with tamoxifen (PROACT) as neoadjuvant therapy in postmenopausal women with hormone receptor-positive breast cancer. *Eur J Cancer Suppl* 2: 69
66 Smith I, Cataliotti L (2004) Anastrozole *versus* tamoxifen as neoadjuvant therapy for oestrogen receptor-positive breast cancer in postmenopausal women: the IMPACT and PROACT trials. *Eur J Cancer Suppl* 2: 69
67 Cuzick J, Forbes J, Edwards R, Baum M, Cawthorn S, Coates A, Hamed A, Howell A, Powles T (2002) First results from the International Breast Cancer Intervention Study (IBIS-I): a randomised prevention trial. *Lancet* 360: 817–824
68 Fisher B, Costantino JP, Wickerham DL, Redmond CK, Kavanah M, Cronin WM, Vogel V, Robidoux A, Dimitrov N, Atkins J et al. (1998) Tamoxifen for prevention of breast cancer: report of the National Surgical Adjuvant Breast and Bowel Project P-1 Study. *J Natl Cancer Inst* 90: 1371–1388
69 Cuzick J, Powles T, Veronesi U, Forbes J, Edwards R, Ashley S, Boyle P (2003) Overview of the main outcomes in breast-cancer prevention trials. *Lancet* 361: 296–300
70 Cuzick J (2003) Aromatase inhibitors in prevention – data from the ATAC (Arimidex, Tamoxifen Alone or in Combination) trial and the design of IBIS-II (the second International Breast Cancer Intervention Study). *Recent Results Cancer Res* 163: 96–103
71 Early Breast Cancer Trialists' Collaborative Group (1998) Tamoxifen for early breast cancer: an overview of the randomised trials. *Lancet* 351: 1451–1467
72 Sawada S, Sato K (2003) Effect of anastrozole and tamoxifen on serum lipid levels in Japanese postmenopausal women with early breast cancer. *Breast Cancer Res Treat* 82 (suppl. 1): S31–S32
73 Wojtacki J, Lesniewski-Kmak K, Piotrowska M, Rolka-Stempniewicz G, Kubik M, Wroblewska M (2004) Effect of anastrozole on serum levels of apolipoprotein A-I and B in patients with early breast cancer: additional data on lack of atherogenic properties. *Breast Cancer Res Treat* 88: S238
74 Wojtacki J, Lesniewski-Kmak K, Pawlak W, Nowicka E (2004) Anastrozole therapy and lipid profile: an update. *Eur J Cancer Suppl* 2: 142
75 Wojtacki J, Lesniewski-Kmak K, Kruszewski WJ (2002) Anastrozole therapy does not compromise lipid metabolism in breast cancer patients previously treated with tamoxifen. *Breast Cancer Res Treat* 76 (suppl. 1): S75
76 Hulley S, Furberg C, Barrett-Connor E, Cauley J, Grady D, Haskell W, Knopp R, Lowery M, Satterfield S, Schrott H et al. (2002) Noncardiovascular disease outcomes during 6.8 years of hormone therapy: Heart and Estrogen/progestin Replacement Study follow-up (HERS II). *JAMA* 288: 58–66
77 Ismail AA, Pye SR, Cockerill WC, Lunt M, Silman AJ, Reeve J, Banzer D, Benevolenskaya LI, Bhalla A, Bruges AJ et al. (2002) Incidence of limb fracture across Europe: results from the European Prospective Osteoporosis Study. *Osteoporos Int* 13: 565–571
78 van Staa TP, Dennison EM, Leufkens HG, Cooper C (2001) Epidemiology of fractures in England and Wales. *Bone* 29: 517–522
79 Writing Group for the Women's Health Initiative (WHI) Investigators (2002) Risks and benefits of estrogen and progestin in healthy postmenopausal women. Principal results from the Women's Health Initiative randomized trial. *J Am Med Assoc* 288: 321–333
80 Winer EP, Hudis C, Burstein HJ, Wolff AC, Pritchard KI, Ingle JN, Chlebowski RT, Gelber R, Edge SB, Gralow J et al. (2005) American Society of Clinical Oncology technology assessment on the use of aromatase inhibitors as adjuvant therapy for postmenopausal women with hormone receptor–positive breast cancer: status report 2004. *J Clin Oncol* 23: 619–629
81 Saphner T, Tormey DC, Gray R (1996) Annual hazard rates of recurrence for breast cancer after primary therapy. *J Clin Oncol* 14: 2738–2746

Aromatase Inhibitors
Edited by B.J.A. Furr



# The third-generation aromatase inhibitors: a clinical overview

Aman Buzdar

*Department of Breast Medical Oncology, The University of Texas M.D. Anderson Cancer Center, 1515 Holcombe Blvd 1354, Houston, TX 77030-4009, USA*

## Introduction

In the USA it is estimated that breast cancer will account for approximately 32% of all new cases of cancer in 2005 [1]. Although treatment has improved and death rates have declined in recent years, breast cancer still accounts for approximately 15% of all cancer deaths in women [1].

Oestrogen is the principal hormone involved in the development of breast cancer. Endocrine agents have, therefore, been designed to block the supply of oestrogen to the breast tumour, either by inhibiting the production of oestrogen or by blocking its action at the oestrogen receptor. For more than 30 years, tamoxifen, an antioestrogen that inhibits the activity of oestradiol at its receptor, has been the mainstay of hormonal therapy for all stages of breast cancer in postmenopausal women. However, tamoxifen does not completely block the activity of the oestrogen receptor, and the remaining partial oestrogen agonist activity is thought to be responsible for some of its unfavourable side effects, such as an increased risk of endometrial cancer [2, 3] and thromboembolic events [4]. In addition, the development of resistance to tamoxifen is a significant problem in breast cancer treatment, and has led to the development of alternative endocrine agents that extend the treatment options for women with hormone-sensitive breast cancer.

Aromatase inhibitors (AIs) act by inhibiting the enzyme aromatase, which catalyses the conversion of androgens to oestrogens, the main source of oestrogens in postmenopausal women. The non-steroidal AI, aminoglutethimide, was the first AI to be introduced, in the late 1970s, for the second-line treatment of advanced breast cancer. However, despite proven efficacy in this setting, its widespread use was limited by its lack of selectivity for aromatase and the resulting toxicity, which meant that concomitant corticosteroid supplementation was necessary [5]. Formestane, a steroidal AI, then became available in 1993; this was more selective and therefore had fewer side effects compared with aminoglutethimide but had to be administered by twice-monthly intramuscular injection [6]. More recently, in the mid-to-late 1990s, the third-generation AIs, anastrozole, letrozole and exemestane, have become available for the treatment of postmenopausal women with advanced and early breast cancer. Anastrozole has the broadest licensing of the third-generation AIs and is approved across the adjuvant and advanced settings. Letrozole is approved as extended adjuvant therapy for women who have already received 5 years' tamoxifen and as first- or second-line treatment of advanced disease. Exemestane is the latest of these third-generation AIs to be approved for advanced breast cancer that has progressed following tamoxifen therapy and does not have a licence in the adjuvant setting.

This chapter summarizes the pharmacology and pharmacokinetics of anastrozole, letrozole, and exemestane and their key efficacy data in breast cancer, from advanced disease in which the AIs were first established as the treatment of choice, to the adjuvant setting, the preoperative setting, and lastly their potential in chemoprevention. Finally, the overall tolerability profiles of the third-generation AIs are reviewed.

## Third-generation AIs

### *Pharmacology and pharmacokinetics*

Anastrozole and letrozole are reversible, imidazole-based, non-steroidal AIs, whereas exemestane is an irreversible steroidal AI (Fig. 1). Although the AIs are often referred to as a class of agents, it is unknown whether the three available third-generation AIs are interchangeable in clinical practice [7].

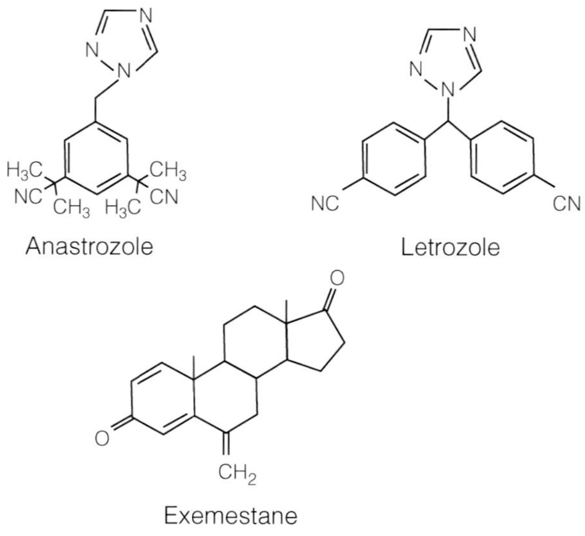


Figure 1. Chemical structures of anastrozole, letrozole and exemestane.

There are several differences between anastrozole, letrozole and exemestane in terms of pharmacokinetics, and effects on lipid profiles and steroidogenesis [8]. The non-steroidal AIs anastrozole and letrozole compete with the endogenous ligands androstenedione and testosterone for the active site of the aromatase enzyme, where they reversibly bond to exclude both ligands and oxygen from the enzyme. In contrast, the steroidal AI exemestane competes with the endogenous ligands for the active site, where it is metabolized to intermediates that bind irreversibly to the active site. Nevertheless, the aromatase enzyme is capable of rapid regeneration so whether its inhibition is reversible or not may have little clinical relevance. The steroidal nature of exemestane does, however, mean that it is associated with androgenic, progestogenic and even some oestrogenic effects; these effects do not occur with anastrozole and letrozole [8]. The AIs also appear to have different effects on plasma lipids and adrenal steroidogenesis [8]. Whereas treatment with anastrozole does not appear to change lipid profiles markedly, statistically significant changes in lipid profiles have been reported with letrozole and exemestane [8]. With respect to their effects on adrenal steroidogenesis, no changes in cortisol or aldosterone levels have been reported with anastrozole or exemestane; however, significant changes have been reported with letrozole [8]. The long-term clinical significance of the differences between these AIs is unknown.

All three third-generation AIs provide potent aromatase inhibition and oestrogen suppression. In the only comparative study of these AIs, a small crossover study in which 12 postmenopausal women with advanced breast cancer received 6 weeks of anastrozole treatment followed by 6 weeks of letrozole or *vice versa*, there was no difference in the level of oestradiol suppression produced by anastrozole and letrozole (85 and 88% reduction, respectively), although letrozole reduced oestrone and oestrone sulfate levels to a significantly greater extent than anastrozole (reduced by 81 *versus* 84% and 94 *versus* 98%, respectively) [9]. However, small differences in oestrogen suppression between the third-generation AIs do not appear to lead to clinically significant differences in overall efficacy [10].

### *Second-line therapy for advanced breast cancer*

A number of phase III, double-blind, randomized studies have evaluated the efficacy and safety of anastrozole [11–13], letrozole [14, 15] and exemestane [16] compared with megestrol acetate in postmenopausal women with advanced breast cancer that is hormone receptor-positive or of unknown hormone receptor status and which has progressed following treatment with tamoxifen or antioestrogens. A summary of the efficacy results of these trials is shown in Table 1. Two trials of identical design were conducted with anastrozole [12, 13], the results of which were presented in prospectively planned combined analyses [11, 17]; two trials were conducted with letrozole [14, 15];

Table 1. Anastrozole, letrozole and exemestane *vs* megestrol acetate as second-line therapy for advanced breast cancer

| Study and reference | No. of patients randomized | Treatment | Follow-up | Summary of efficacy results |
|---|---|---|---|---|
| Anastrozole | | | | |
| Buzdar et al. [11, 17] | 764 | Anastrozole 1 mg/day, anastrozole 10 mg/day or megestrol acetate 160 mg/day | Median 31 months | ORR, 13, 13 and 12% (NS). TTP, HR 0.94 (97.5% CI 0.76–1.16; NS) for 1 mg *vs* MA; HR 0.91 (97.5% CI 0.73–1.12; NS) for 10 mg *vs* MA. OS, HR 0.78 (97.5% CI 0.6–1.0; $P < 0.025$) for 1 mg *vs* MA; HR 0.83 (97.5% CI 0.64–1.1; NS) for 10 mg *vs* MA |
| Letrozole | | | | |
| Dombernowsky et al. [15] | 551 | Letrozole 0.5 mg/day, letrozole 2.5 mg/day or megestrol acetate 160 mg/day | Median 33 months (tumour response and safety) or 45 months (survival) | ORR, 13 *vs* 24% ($P = 0.04$ *vs* MA and $P = 0.004$ *vs* 0.5 mg) *vs* 16%. TTP, HR 1.04 (95% CI 0.81–1.32; NS) for 0.5 mg *vs* MA; HR 0.80 (95% CI 0.62–1.02; NS) for 2.5 mg *vs* MA. OS, HR 1.12 (95% CI 0.87–1.44; NS) for 0.5 mg *vs* MA; HR 0.82 (95% CI 0.63–1.09; NS) for 2.5 mg *vs* MA |
| Buzdar et al. [14] | 602 | Letrozole 0.5 mg/day, letrozole 2.5 mg/day or megestrol acetate 160 mg/day | 30-month enrolment period, followed by 18 months of follow-up (tumour response and safety) or 37 months of follow-up (survival) | ORR, 21, 16 and 15% (NS). TTP, HR 0.80 (95% CI 0.64–0.99; $P = 0.044$) for 0.5 mg *vs* MA; HR 0.99 (95% CI 0.79–1.23; NS) for 2.5 mg *vs* MA. OS, HR 0.79 (95% CI 0.62–1.00; NS) for 0.5 mg *vs* MA; HR 0.92 (95% CI 0.73–1.17; NS) for 2.5 mg *vs* MA |
| Exemestane | | | | |
| Kaufmann et al. [16] | 769 | Exemestane 25 mg/day or megestrol acetate 160 mg/day | Median 11 months | ORR, 15 *vs* 12% (NS). TTP, 5 *vs* 4 months ($P = 0.037$). OS, median not yet reached *vs* 28 months ($P = 0.039$) |

CI, confidence interval; HR, hazard ratio; MA, megestrol acetate; NS, non significant; ORR, objective response rate; OS, overall survival; TTP, time to progression.

and one trial was conducted with exemestane [16]. Anastrozole, letrozole and exemestane each demonstrated significant clinical benefit compared with megestrol acetate (Tab. 1). Anastrozole was the only AI to demonstrate clearly a significant survival benefit compared with megestrol acetate based on mature data with prolonged follow-up. Initial results for exemestane demonstrated a survival advantage for the AI but an updated analysis has yet to be published. Of the two letrozole studies, the first showed a dose response for letrozole with a statistically significantly higher objective response rate for letrozole 2.5 mg compared with megestrol acetate [15], whereas the second did not replicate the statistical superiority of letrozole 2.5 mg *versus* megestrol acetate although letrozole 0.5 mg did show clinical benefit [14].

There has been only one head-to-head study directly comparing third-generation AIs as second-line treatment for advanced disease [18]. In this open-label, randomized study comparing treatment with anastrozole and letrozole in 713 patients with advanced breast cancer that was hormone receptor-positive (48% of the total population) or of unknown hormone receptor status, no significant difference was found in the primary efficacy endpoint of time to progression or in the secondary endpoint of overall survival. A significantly higher objective response rate occurred with letrozole compared with anastrozole (19 *versus* 12%; $P = 0.013$) but there was no significant difference in the clinically relevant target population of patients known to be hormone receptor-positive (17% for both treatments) [18].

The proven efficacy of the third-generation AIs, together with their significant tolerability advantages compared with megestrol acetate (see below), led rapidly to their acceptance as the first-choice endocrine therapy for the second-line treatment of advanced breast cancer.

### *First-line therapy for advanced breast cancer*

Following successful results in the second-line setting, a number of phase III trials investigated the efficacy and safety of the third-generation AIs as first-line therapy for advanced breast cancer that was hormone receptor-positive or of unknown hormone receptor status in postmenopausal women. Key efficacy data are shown in Table 2, including combined data from two studies comparing anastrozole and tamoxifen that were prospectively designed to allow for combined analyses [19–22], data from a smaller independent study comparing anastrozole with tamoxifen [23], data from a study comparing letrozole and tamoxifen [24, 25] and preliminary data from a comparison of exemestane and tamoxifen [26]. The phase III trials of letrozole and exemestane compared with tamoxifen were prospectively designed to test superiority of the AI, unlike the two large anastrozole trials, which were designed to show equivalence in the primary endpoints.

The third-generation AIs were all shown to have at least equivalent or superior efficacy to tamoxifen as first-line treatment of postmenopausal women

Table 2. Anastrozole, letrozole and exemestane *vs* tamoxifen as first-line therapy for advanced breast cancer

| Study and reference | No. of patients randomized | Treatment | Median follow-up | Summary of efficacy results |
|---|---|---|---|---|
| Anastrozole | | | | |
| Bonneterre et al. [19], Nabholtz et al. [21] | 1021 | Anastrozole 1 mg/day or tamoxifen 20 mg/day | 18 months (TTP and tumour response) or 44 months (survival and tolerability) | TTP, 8.5 *vs* 7.0 months; HR 1.13 (lower 95% CI 1.00). For patients with HR+ tumours: TTP, 10.7 *vs* 6.4 months; $P = 0.022$; ORR, 29 *vs* 27%; OS, median 39 *vs* 40 months (HR 0.97; lower 95% CI 0.84). For patients with HR+ tumours: survival, median 41 months in both arms (HR 1.00; lower 95% CI 0.83) |
| Milla-Santos et al. [23] | 238 | Anastrozole 1 mg/day or tamoxifen 40 mg/day | 13 months | TTP, NA. ORR, 36 *vs* 26% (NS). OS, median 17.4 *vs* 16.0 months (HR 0.64; 95% CI 0.47–0.86; $P = 0.003$) |
| Letrozole | | | | |
| Mouridsen et al. [24, 25] | 916 | Letrozole 2.5 mg/day or tamoxifen 20 mg/day | 32 months | TTP, 9.4 *vs* 6.0 months; HR 0.72 ($P < 0.0001$). ORR, 32 *vs* 21% (OR 1.78; $P = 0.0002$). OS, 34 *vs* 30 months (NS) |
| Exemestane | | | | |
| Paridaens et al. [26] | 382 | Exemestane 25 mg/day or tamoxifen 20 mg/day | 29 months | PFS, 9.9 *vs* 5.8 months; HR 0.84 (95% CI 0.67–1.05; NS). ORR, 46 *vs* 31% (OR 1.85; 95% CI 1.21–2.82; $P = 0.005$). OS, HR 1.04 (95% CI 0.76–1.41; NS) |

CI, confidence intervals; HR, hazard ratio; HR+, hormone receptor-positive; NA, not available; NS, non significant; OR, odds ratio; ORR, objective response rate; OS, overall survival; PFS, progression-free survival; TTP, time to progression with advanced breast cancer of hormone receptor-positive or unknown hormone receptor status. Although anastrozole improved time to progression compared with tamoxifen in the total population of the combined anastrozole trials, the difference was statistically significant only in the hormone receptor-positive population (60% of the total population) [19]. Letrozole significantly improved time to progression and objective tumour response in the total population (66% of whom were hormone receptor-positive) compared with tamoxifen [25]. Although truncated log-rank tests at 6-month intervals showed nominally statistically significant differences in survival in favour of the randomized letrozole arm between 6 months and 2 years, these were non-protocol-specified, retrospectively planned analyses, and the prospectively planned final analysis of overall survival, at a median follow-up of 32 months, found no significant difference in overall survival between letrozole and tamoxifen [24]. Exemestane significantly improved objective tumour response in the total population (~87% of whom were hormone receptor-positive) compared with tamoxifen but although it improved progression-free survival, this did not reach statistical significance [26]. The only significant difference in survival was in the independent anastrozole study, in which anastrozole significantly prolonged median survival compared with tamoxifen in postmenopausal women with oestrogen receptor-positive tumours [23]. The patient population in this study was distinct from that in the other first-line studies in that all of the patients were oestrogen receptor-positive, none had received prior hormonal adjuvant therapy and the majority received only palliative care after relapse; thus, it is perhaps the ideal population in which to detect a survival difference between an AI and tamoxifen without the confounding influence of other endocrine treatment. However, it is worth noting that to conduct such a study is not ethical due to the number of therapeutic agents that are available to provide disease control and palliation in the advanced setting.

### *Adjuvant therapy for early breast cancer*

#### *Newly diagnosed women*

Anastrozole is currently the only AI with data in the primary adjuvant setting over the full 5-year recommended treatment period [27] (Tab. 3). When the ATAC (Arimidex, Tamoxifen, Alone or in Combination) trial was started, women with negative or unknown hormone-receptor status were still thought to derive some benefit from adjuvant therapy with a hormonal agent and were, therefore, included in the ATAC trial. Consequently, only 84% of this trial population were confirmed as hormone receptor-positive. Initial primary adjuvant results comparing letrozole with tamoxifen from the BIG (Breast International Group) 1-98 trial have been presented recently (Tab. 3) [28]. Positive hormone-receptor status was an eligibility criterion for the BIG 1-98 trial. A primary adjuvant trial, TEAM (Tamoxifen-Exemestane Adjuvant Multicenter), comparing exemestane with tamoxifen, is also in progress.

Table 3. Anastrozole, letrozole and exemestane as adjuvant therapy for early breast cancer

| Study and reference | No. of patients randomized | Treatment | Median follow-up | Summary of efficacy results |
|---|---|---|---|---|
| Primary adjuvant therapy | | | | |
| Anastrozole | | | | |
| ATAC Trialists' Group [27] | 9366 | Anastrozole 1 mg/day, tamoxifen 20 mg/day or anastrozole 1 mg/day plus tamoxifen 20 mg/day[a] | 68 months | DFS, HR 0.87 (95% CI 0.78–0.97; $P = 0.01$); in HR+ patients, HR 0.83 (95% CI 0.73–0.94; $P = 0.005$). TTR, HR 0.79 (95% CI 0.70–0.90; $P = 0.0005$); in HR+ patients, HR 0.74 (95% CI 0.64–0.87; $P = 0.0002$). TTDR, HR 0.86 (95% CI 0.74–0.99; $P = 0.04$); in HR+ patients, HR 0.84 (95% CI 0.70–1.00; NS). OS, HR 0.97 (95% CI 0.85–1.12; NS). Contralateral breast cancer: OR, 0.58 (95% CI 0.38–0.88; $P = 0.01$) |
| Letrozole | | | | |
| BIG 1-98 Collaborative Group [28] | 8028[b] | Letrozole 2.5 mg/day, tamoxifen 20 mg/day, letrozole for 2 years followed by tamoxifen for 3 years or tamoxifen for 2 years followed by letrozole for 3 years | 26 months | DFS, HR 0.81 (95% CI 0.70–0.93; $P = 0.003$)[c]. TTR, HR 0.72 (95% CI 0.61–0.86; $P = 0.0002$). TTDR, HR 0.73 (95% CI 0.60–0.88; $P = 0.0012$). OS, HR 0.86 (95% CI 0.70–1.06; $P = 0.16$) |
| After 2–3 years of tamoxifen | | | | |
| Anastrozole | | | | |
| Jakesz [31] | 3224 | Anastrozole 1 mg/day for 3 years or tamoxifen 20 mg/day for 3 years, both after 2 years' prior tamoxifen | 28 months | EFS, HR 0.60 (95% CI 0.44–0.81; $P = 0.0009$). DRFS, HR 0.61 (95% CI 0.42–0.87; $P = 0.0067$). OS, HR 0.76 (95% CI 0.52–1.12; NS) |

(Continued on next page)

Table 3. (Continued)

| Study and reference | No. of patients randomized | Treatment | Median follow-up | Summary of efficacy results |
|---|---|---|---|---|
| ITA [32] | 448 | Anastrozole 1 mg/day or tamoxifen 20 mg/day after 2–3 years' prior tamoxifen (total duration of treatment 5 years) | 36 months | EFS, HR 0.35 (95% CI 0.20–0.63; $P = 0.0002$). RFS, HR 0.35 (95% CI 0.18–0.68; $P = 0.001$). DRFS, HR 0.49 (95% CI 0.22–1.05; NS) |
| Exemestane | | | | |
| IES [33] | 4742 | Exemestane 25 mg/day or tamoxifen 20 mg/day after 2–3 years' prior tamoxifen (total duration of treatment 5 years) | 31 months | DFS, HR 0.68 (95% CI 0.56–0.82; $P = 0.00005$). DRFS, HR 0.66 (95% CI 0.52–0.83; $P = 0.0004$). OS, HR 0.88 (95% CI 0.67–1.16; NS). Contralateral breast cancer, HR 0.44 (95% CI 0.20–0.98; $P = 0.04$) |
| After 5 years of tamoxifen (extended adjuvant therapy) | | | | |
| Letrozole | | | | |
| MA-17 [36, 44] | 5187 | Letrozole 2.5 mg/day or placebo after 5 years' prior tamoxifen | 30 months | DFS, HR 0.58 ($P = 0.00004$). DDFS, HR 0.60 ($P = 0.002$). OS, NS. Reduced risk of contralateral breast cancer by 37.5% |

[a] The combination arm was closed because of low efficacy after the analysis at 47 months' median follow-up; results are presented for the monotherapy arms only.
[b] 1835 randomized to arms comparing letrozole with tamoxifen, followed by 6193 randomized to all four arms including crossover arms.
[c] The BIG 1-98 definition of DFS includes non-breast cancer primary cancers as events; these were not included in the DFS endpoint in the ATAC trial.
CI, confidence interval; DDFS, distant disease-free survival; DFS, disease-free survival; DRFS, distant recurrence-free survival; EFS, event-free survival; HR, hazard ratio; HR+, hormone receptor-positive; OR, odds ratio; OS, overall survival; RFS, recurrence-free survival; TTDR, time to distant recurrence; TTR, time to recurrence.

Data from the completed treatment analysis of the ATAC trial, at a median follow-up of 68 months ($n = 9366$), have confirmed the findings of earlier analyses [29, 30] showing that anastrozole significantly prolongs disease-free survival, time to recurrence and time to distant recurrence, and significantly reduces contralateral breast cancers (Fig. 2, Tab. 3) [27]. Initial data from the BIG 1-98 trial, at a median follow-up of 26 months ($n = 8028$), have also shown that letrozole significantly prolongs disease-free survival, time to recurrence and time to distant recurrence [28]. Neither study has yet shown a survival advantage for the AI over tamoxifen. As anastrozole is the only AI with long-term efficacy and tolerability data and an established risk/benefit profile in the primary adjuvant setting, current evidence suggests that anastrozole should be the preferred initial treatment for postmenopausal women with localized hormone receptor-positive breast cancer; it is currently the only AI approved for this indication [27].

*Women already receiving adjuvant tamoxifen therapy*

Women who are already part way through a course of adjuvant therapy with tamoxifen may benefit from switching to an AI. This approach has been investigated in several switching trials in which patients who had already received 2–3 years' adjuvant tamoxifen were randomized to continued tamoxifen or to an AI (Tab. 3). There have been three such switching trials with anastrozole: the ABCSG (Austrian Breast and Colorectal Cancer Study Group) 8 and ARNO (Arimidex-Nolvadex) 95 trials, which included postmenopausal women with hormone receptor-positive disease who had already received 2

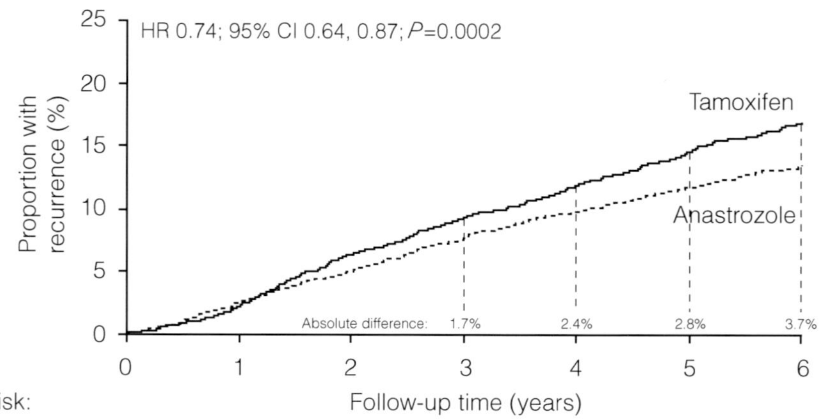


| At risk: | 0 | 1 | 2 | 3 | 4 | 5 | 6 |
|---|---|---|---|---|---|---|---|
| Anastrozole | 2618 | 2540 | 2448 | 2355 | 2268 | 2014 | 830 |
| Tamoxifen | 2598 | 2516 | 2398 | 2304 | 2189 | 1932 | 774 |

Figure 2. Anastrozole as primary adjuvant therapy for early breast cancer: time to recurrence in patients with hormone receptor-positive tumours. Reprinted from [27] with permission from Elsevier. CI, confidence interval; HR, hazard ratio.

years' tamoxifen and whose results have been presented in a prospectively planned combined analysis ($n = 3224$) [31]; and the smaller ITA (Italian Tamoxifen Anastrozole) trial, which included postmenopausal women with oestrogen receptor-positive disease who had already received 2–3 years of tamoxifen ($n = 448$) [32]. A similar randomized study with exemestane, the IES (International Exemestane Study) [33], included postmenopausal women with oestrogen receptor-positive early breast cancer who had already received 2–3 years tamoxifen treatment ($n = 4742$).

The results of these studies indicate that switching to either anastrozole or exemestane after 2–3 years of tamoxifen for the remainder of the standard 5-year adjuvant treatment period significantly prolongs event-free survival and distant recurrence-free survival compared with continued tamoxifen treatment, although there were no significant differences in overall survival at the time of the analyses (Tab. 3). The initial analysis of the ABCSG 8 trial, at a median follow-up of 28 months, met the stopping boundary for event-free survival; therefore, it was recommended that accrual was terminated and the patients informed of the results. The second interim analysis of the IES trial also met the stopping boundary for the trial and the efficacy results were reported at a median follow-up of 31 months. Overall, the results of these studies suggest that postmenopausal breast cancer patients who are already receiving adjuvant tamoxifen should switch to anastrozole or exemestane after 2–3 years of tamoxifen; exemestane is not currently licensed for adjuvant therapy.

Although those patients who are already receiving adjuvant tamoxifen may benefit from switching to an AI, evidence suggests that the most appropriate therapy for newly diagnosed patients is to start with the most effective therapy at the earliest opportunity. The risk of breast cancer recurrence is highest during the first 5 years post-surgery, peaking at 2–3 years [34]. The lower rates of recurrence with anastrozole, particularly within the first 3 years post-surgery, and the lower incidence of adverse events and treatment withdrawals compared with tamoxifen demonstrated in the ATAC completed treatment analysis, justify starting treatment with anastrozole rather than starting treatment with tamoxifen with the intention of switching to an AI.

*Women who have completed 5 years of adjuvant tamoxifen*

Adjuvant treatment with tamoxifen is only recommended for 5 years; studies have shown that there is no additional benefit from tamoxifen administered beyond 5 years [4, 35]. However, these women could still benefit from continued endocrine therapy with an alternative agent. The MA-17 trial [36] investigated whether postmenopausal women who had already completed 4.5–6 years of adjuvant tamoxifen would benefit from further treatment with letrozole (Tab. 3). This study showed that letrozole significantly prolonged disease-free survival and distant disease-free survival, and reduced the risk of contralateral breast cancer compared with placebo. There was no significant difference in overall survival. These results led to the recommended termination of the trial and communication of the results to the participants. Although these results show that in postmenopausal women letrozole therapy after the completion of standard tamoxifen treatment significantly improves disease-free survival, early discontinuation of the trial means that the optimal duration of the treatment, and long-term tolerability, remain undefined.

*Summary of adjuvant therapy studies*

Based on the updated analysis of the ATAC trial at a median follow-up of 47 months [30], and on data from the ITA, IES and MA-17 trials, the Anerican Society of Clinical Oncology (ASCO) technology assessment, conducted in 2004, concluded that 'optimal adjuvant hormonal therapy for a postmenopausal woman with receptor-positive breast cancer includes an AI as initial therapy or after treatment with tamoxifen' to reduce the risk of tumour recurrence [7]. This statement has since been further corroborated by the results of the completed treatment analysis of the ATAC trial at a median follow-up of 68 months, the first analysis of the BIG 1-98 trial, and the combined analysis of the ABCSG 8/ARNO 95 trials.

Anastrozole is currently the only AI with long-term efficacy and tolerability data, provided by the completed treatment analysis of the ATAC trial. The other studies are limited by the immaturity of the data. In addition, although these trials show that patients clearly benefit from treatment with an AI as initial adjuvant therapy or after prior tamoxifen, they do not indicate the optimum sequencing of endocrine therapy. As these trials are of different designs and include different patient populations, it is inappropriate to make any cross-trial comparisons of efficacy. Continuing clinical trials should help to define the optimal timing, duration and sequencing of AI therapy, and potential differences between anastrozole, letrozole and exemestane. The ASCO technology assessment noted that 'it is unknown if the three available drugs are interchangeable in clinical practice' and the panel favoured 'using the AI that has been studied in the setting most closely approximating any individual patient's clinical circumstance' [7].

## *Preoperative therapy for early breast cancer*

Both anastrozole and letrozole have been investigated as preoperative therapy in randomized direct comparison with tamoxifen. Two randomized, double-blind studies – PROACT (Preoperative Arimidex (anastrozole) Compared with Tamoxifen) and IMPACT (Immediate Preoperative Arimidex, tamoxifen or Combined with Tamoxifen) – compared preoperative treatment with anastrozole and tamoxifen in postmenopausal women with large operable or inoperable (including locally advanced), hormone receptor-positive breast cancer [37, 38]. Data from these studies were combined in a prospectively planned analysis [39]. Another study compared preoperative treatment with letrozole and tamoxifen in postmenopausal women with hormone receptor-positive breast cancer that was considered inoperable or not eligible for breast-conserving surgery [40]. The efficacy data for the clinically relevant populations of patients with inoperable disease or who required a mastectomy at trial entry are shown in Table 4.

Both anastrozole and letrozole provided effective preoperative treatment, producing clinically beneficial reductions in tumour volume to enable breast-conserving surgery in patients previously only eligible for mastectomy. Therefore, these AIs are a beneficial preoperative option for postmenopausal women with early stage breast cancer who have disease that is considered inoperable or not eligible for breast-conserving surgery, or for those women who do not wish to or are unable to undergo immediate surgery or preoperative chemotherapy.

### *Chemoprevention*

Five years of treatment with adjuvant tamoxifen reduces the risk of contralateral breast cancer by approximately 50% in women with oestrogen receptor-positive tumours compared with no tamoxifen [2], and a meta-analysis of prevention studies has shown that tamoxifen reduces the incidence of breast cancer by 38% in women at high risk compared with placebo [41]. Currently, tamoxifen is the only hormonal therapy approved by the US Food and Drug Administration for the prevention of breast cancer in women considered high risk; however, the AIs have the potential to prevent even more patients at high risk of breast cancer from developing tumours.

Anastrozole [27], exemestane [33] and letrozole [36] have all been shown to reduce significantly the incidence of contralateral breast cancer in postmenopausal women with early-stage breast cancer compared with tamoxifen. Thus prophylactic treatment with these AIs might be more effective than tamoxifen in preventing tumours in women at high risk of breast cancer.

Phase III trials are in progress to test the efficacy of the third-generation AIs in the prevention of breast cancer, including the IBIS (International Breast Cancer Intervention Study) II of anastrozole *versus* tamoxifen, and the NCIC CTG (National Cancer Institute of Canada Clinical Trials Group) MAP.3 trial of exemestane *versus* placebo.

### *Tolerability*

Anastrozole, letrozole and exemestane differ in their chemical structures, pharmacokinetics, effects on lipid profiles and steroidogenesis, and perhaps the degree to which they suppress aromatase activity [8]. The clinical significance of these differences is unknown, but the ASCO panel in 2002 noted that ‘closely related agents with similar mechanisms of action may have different toxicity profiles’. Here, we review the similarities and potential differences between

Table 4. Anastrozole and letrozole as preoperative therapy

| Study and reference | Patient population | No. of patients randomized | Treatment | Summary of efficacy results |
|---|---|---|---|---|
| Anastrozole | | | | |
| Smith and Cataliotti [39] | Primary HR+ (the PROACT trial) or ER+ (the IMPACT trial) breast cancer, considered inoperable or not eligible for breast-conserving surgery[a] | 344 | Anastrozole 1 mg/day or tamoxifen 20 mg/day for 12 weeks prior to surgery | Calliper response, 47 *vs* 35% (OR 1.65; 95% CI 1.06–2.56; $P = 0.026$). Ultrasound response, 36 *vs* 26% (OR 1.60; 95% CI 1.00–2.55; $P = 0.048$). Breast-conserving surgery, 43 *vs* 31% (OR 1.70; 95% CI 1.09–2.66; $P = 0.019$) |
| Letrozole | | | | |
| Eiermann et al. [40] | Primary, untreated HR+ breast cancer, considered inoperable or not eligible for breast-conserving surgery | 337 | Letrozole 2.5 mg/day or tamoxifen 20 mg/day for 4 months prior to surgery | Objective tumour response, 55 *vs* 36% ($P < 0.001$; OR 2.23; 95% CI 1.43–3.50; $P = 0.0005$). Ultrasound response, 35 *vs* 25% ($P = 0.042$). Mammographic response, 34 *vs* 16% ($P < 0.001$). Breast-conserving surgery, 45 *vs* 35%; |
| $P = 0.022$ | | | | |

[a] Results are presented here only for those patients who were considered inoperable or not eligible for breast-conserving surgery.
CI, confidence intervals; ER+, oestrogen receptor-positive; HR+, hormone receptor-positive; OR, odds ratio.

the third-generation AIs from clinical trial data to date. Anastrozole has the most mature adverse-event data of the third-generation AIs: it is the only AI with long-term tolerability data up to and beyond 5 years of follow-up. We will, therefore, compare the adverse-event data for anastrozole in the ATAC trial with those for letrozole and exemestane in comparative studies with tamoxifen. However, these trials are of different designs and include different patient populations; therefore, any cross-trial comparisons should be interpreted with caution.

In the ATAC trial, anastrozole was associated with significant reductions in the incidence of endometrial cancer, thromboembolic events, ischaemic cerebrovascular events, vaginal bleeding, hot flushes and vaginal discharge compared with tamoxifen [27]. Similarly, in the IES trial, exemestane was associated with significant reductions in the incidence of thromboembolic disease, vaginal bleeding and gynaecological symptoms [33]. In this trial, exemestane was also associated with a significantly reduced incidence of muscle cramps. Endometrial cancer developed in fewer patients in the exemestane group than in the tamoxifen group but the difference was not statistically significant. In the first analysis of the BIG 1-98 trial, letrozole was associated with a reduced incidence of thromboembolic events and vaginal bleeding (statistical significance not available) [28]. Again, endometrial cancer developed in fewer patients in the letrozole group than in the tamoxifen group, although this did not reach statistical significance.

In the ATAC trial, tamoxifen was associated with significant reductions in the incidence of arthralgia and fractures, although there was no significant difference between anastrozole and tamoxifen for fractures of the hip – the fracture type with the highest morbidity and mortality [27]. Exemestane has also been associated with an increased risk of osteoporosis ($P = 0.05$) and an increased incidence of fractures compared with tamoxifen, although the difference was not statistically significant [33]. Letrozole was associated with a statistically significantly increased incidence of fractures in the BIG 1-98 trial [28]. In recognition of the potential effect of AIs on bone mineral density and subsequent fracture risk, bone density testing and, if indicated, appropriate treatment with bisphosphonates, have been recommended for postmenopausal women receiving AIs for breast cancer [42]. A significantly increased incidence of arthralgia has also been reported for exemestane compared with placebo [33]. In comparison with tamoxifen, exemestane was also associated with significantly increased incidences of visual disturbances and diarrhoea. In the BIG 1-98 trial, letrozole was associated with an increased incidence of hypercholesterolaemia [28]; however, cholesterol levels were not systematically measured in the ATAC and IES trials.

Perhaps of more concern in the BIG 1-98 trial is the increased incidence of ‘other cardiovascular adverse events’ of grade 3–5 (excluding cerebrovascular accidents/transient ischemic attack, and thromboembolic events; 3.6 *versus* 2.5%) and the increased number of cerebrovascular (7 *versus* 1) and cardiovascular (26 *versus* 13) deaths with letrozole compared with tamoxifen [28].

In comparison, there is no evidence of a cardiovascular safety issue with anastrozole [27]; although ischemic cardiovascular events were reported more frequently with anastrozole relative to tamoxifen in the ATAC trial, there was no significant difference. There was a similar number of cardiovascular deaths in the anastrozole and tamoxifen groups (49 *versus* 46, respectively [43]). The IES trial reported a higher incidence of myocardial infarction with exemestane compared with tamoxifen but the difference was not statistically significant (1.0 *versus* 0.4%).

The tolerability and safety data for anastrozole are mature after more than 5 years of follow-up and show that anastrozole is associated with significantly reduced incidences of endometrial cancer, thromboembolic and cerebrovascular events compared with tamoxifen. At 26 months of follow-up, the BIG 1-98 trial raises concerns about the cardiovascular side effects of letrozole: a longer follow-up is required to determine fully the risk/benefit profile of letrozole in the adjuvant setting.

## Conclusions

The third-generation AIs are established as the endocrine treatment of choice for advanced breast cancer. In the adjuvant setting, the ASCO technology assessment of 2004 favoured 'using the aromatase inhibitor that has been most studied in the setting most closely approximating any individual patient's clinical circumstance' [7]. Thus anastrozole is most frequently the AI of choice as primary adjuvant therapy. If a patient has already received 2–3 years of adjuvant tamoxifen, switching to anastrozole or exemestane may be appropriate, and for those patients who have completed 5 years of adjuvant tamoxifen, extended adjuvant therapy with letrozole is an appropriate treatment choice. Both anastrozole and letrozole also have data supporting their use in the preoperative setting.

Some important questions remain unanswered, including the optimal duration and sequencing of adjuvant treatment with an AI, the long-term toxicities and risks associated with letrozole and exemestane, whether there are any important clinical differences between the third-generation AIs, and the efficacy of the AIs in the chemoprevention of breast cancer. Continuing clinical trials should provide answers to these questions.

There is no doubt that the third-generation AIs now play an important role in the treatment of postmenopausal women with breast cancer. As new trial results become available, physicians and patients will need to reconsider carefully the currently available data applicable to their own particular circumstances when deciding the optimal treatment strategy.

## References


1 American Cancer Society (2005) *Cancer facts and Figures 2005*. http://www.cancer.org/downloads/STT/CAFF2005f4PWSecured.
2 Early Breast Cancer Trialists' Collaborative Group (1998) Tamoxifen for early breast cancer: an overview of the randomised trials. *Lancet* 351: 1451–1467
3 Wysowski DK, Honig S, Beitz J (2002) Uterine sarcoma associated with tamoxifen use. *N Engl J Med* 346: 1832–1833
4 Fisher B, Dignam J, Bryant J, DeCillis A, Wickerham DL, Wolmark N, Costantino J, Redmond C, Fisher ER, Bowman DM, Deschenes L et al. (1996) Five *versus* more than five years of tamoxifen therapy for breast cancer patients with negative lymph nodes and estrogen receptor-positive tumours. *J Natl Cancer Inst* 88: 1529–1542
5 Santen RJ, Worgul TJ, Lipton A, Harvey H, Boucher A, Samojlik E, Wells SA (1982) Aminoglutethimide as treatment of postmenopausal women with advanced breast carcinoma. *Ann Intern Med* 96: 94–101
6 Goss PE, Powles TJ, Dowsett M, Hutchison G, Brodie AM, Gazet JC, Coombes RC (1986) Treatment of advanced postmenopausal breast cancer with an aromatase inhibitor, 4-hydroxyandrostenedione: phase II report. *Cancer Res* 46: 4823–4826
7 Winer EP, Hudis C, Burstein HJ, Wolff AC, Pritchard KI, Ingle JN, Chlebowski RT, Gelber R, Edge SB, Gralow J et al. (2005) American Society of Clinical Oncology technology assessment on the use of aromatase inhibitors as adjuvant therapy for postmenopausal women with hormone receptor–positive breast cancer: status report 2004. *J Clin Oncol* 23: 619–629
8 Buzdar A, Robertson JFR, Eiermann W, Nabholtz JM (2002) An overview of the pharmacology and pharmacokinetics of the newer generation aromatase inhibitors anastrozole, letrozole and exemestane. *Cancer* 95: 2006–2016
9 Geisler J, Haynes B, Anker G, Dowsett M, Lønning PE (2002) Influence of letrozole and anastrozole on total body aromatization and plasma estrogen levels in postmenopausal breast cancer patients evaluated in a randomized, cross-over study. *J Clin Oncol* 20: 751–757
10 Sainsbury R (2004) Aromatase inhibition in the treatment of advanced breast cancer: is there a relationship between potency and clinical efficacy? *Br J Cancer* 90: 1733–1739
11 Buzdar AU, Jonat W, Howell A, Jones SE, Blomqvist CP, Vogel CL, Eiermann W, Wolter JM, Steinberg M, Webster A, Lee D (1998) Anastrozole *versus* megestrol acetate in the treatment of postmenopausal women with advanced breast carcinoma: results of a survival update based on a combined analysis of data from two mature phase III trials. Arimidex Study Group. *Cancer* 83: 1142–1152
12 Buzdar AU, Jones SE, Vogel CL, Wolter J, Plourde P, Webster A (1997) A phase III trial comparing anastrozole (1 and 10 milligrams), a potent and selective aromatase inhibitor, with megestrol acetate in postmenopausal women with advanced breast carcinoma. Arimidex Study Group. *Cancer* 79: 730–739
13 Jonat W, Howell A, Blomqvist C, Eiermann W, Winblad G, Tyrrell C, Mauriac L, Roche H, Lundgren S, Hellmund R, Azab M (1996) A randomised trial comparing two doses of the new selective aromatase inhibitor anastrozole (Arimidex) with megestrol acetate in postmenopausal patients with advanced breast cancer. *Eur J Cancer* 32A: 404–412
14 Buzdar A, Douma J, Davidson N, Elledge R, Morgan M, Smith R, Porter L, Nabholtz J, Xiang X, Brady C (2001) Phase III, multicenter, double-blind, randomized study of letrozole, an aromatase inhibitor, for advanced breast cancer *versus* megestrol acetate. *J Clin Oncol* 19: 3357–3366
15 Dombernowsky P, Smith I, Falkson G, Leonard R, Panasci L, Bellmunt J, Bezwoda W, Gardin G, Gudgeon A, Morgan M et al. (1998) Letrozole, a new oral aromatase inhibitor for advanced breast cancer: double-blind randomized trial showing a dose effect and improved efficacy and tolerability compared with megestrol acetate. *J Clin Oncol* 16: 453–461
16 Kaufmann M, Bajetta E, Dirix LY, Fein LE, Jones SE, Zilembo N, Dugardyn JL, Nasurdi C, Mennel RG, Cervek J et al. (2000) Exemestane is superior to megestrol acetate after tamoxifen failure in postmenopausal women with advanced breast cancer: results of a phase III randomized double-blind trial. The Exemestane Study Group. *J Clin Oncol* 18: 1399–1411
17 Buzdar A, Jonat W, Howell A, Jones SE, Blomqvist C, Vogel CL, Eiermann W, Wolter JM, Azab M, Webster A, Plourde PV (1996) Anastrozole, a potent and selective aromatase inhibitor, *versus* megestrol acetate in postmenopausal women with advanced breast cancer: results of overview analysis of two phase III trials. Arimidex Study Group. *J Clin Oncol* 14: 2000–2011

18 Rose C, Vtoraya O, Pluzanska A, Davidson N, Gershanovich M, Thomas R, Johnson S, Caicedo JJ, Gervasio H, Manikhas G et al. (2003) An open randomised trial of second-line endocrine therapy in advanced breast cancer: comparison of the aromatase inhibitors letrozole and anastrozole. *Eur J Cancer* 39: 2318–2327

19 Bonneterre J, Buzdar A, Nabholtz JM, Robertson JF, Thürlimann B, von Euler M, Sahmoud T, Webster A, Steinberg M, Arimidex Writing Committee, Investigators Committee Members (2001) Anastrozole is superior to tamoxifen as first-line therapy in hormone receptor positive advanced breast carcinoma. *Cancer* 92: 2247–2258

20 Bonneterre J, Thürlimann B, Robertson JF, Krzakowski M, Mauriac L, Koralewski P, Vergote I, Webster A, Steinberg M, von Euler M (2000) Anastrozole *versus* tamoxifen as first-line therapy for advanced breast cancer in 668 postmenopausal women: results of the Tamoxifen or Arimidex Randomized Group Efficacy and Tolerability study. *J Clin Oncol* 18: 3748–3757

21 Nabholtz JM, Bonneterre J, Buzdar A, Robertson JFR, Thürlimann B, for the Arimidex Writing Committee on behalf of the Investigators (2003) Anastrozole (Arimidex™) *versus* tamoxifen as first-line therapy for advanced breast cancer in postmenopausal women: survival analysis and updated safety results. *Eur J Cancer* 39: 1684–1689

22 Nabholtz JM, Buzdar A, Pollak M, Harwin W, Burton G, Mangalik A, Steinberg M, Webster A, von Euler M (2000) Anastrozole is superior to tamoxifen as first-line therapy for advanced breast cancer in postmenopausal women: results of a North American multicenter randomized trial. Arimidex Study Group. *J Clin Oncol* 18: 3758–3767

23 Milla-Santos A, Milla L, Portella J, Rallo L, Pons M, Rodes E, Casanovas J, Puig-Gali M (2003) Anastrozole *versus* tamoxifen as first-line therapy in postmenopausal patients with hormone-dependent advanced breast cancer: a prospective, randomized, phase III study. *Am J Clin Oncol* 26: 317–322

24 Mouridsen H, Gershanovich M, Sun Y, Perez-Carrion R, Boni C, Monnier A, Apffelstaedt J, Smith R, Sleeboom HP, Jaenicke F et al. (2003) Phase III study of letrozole *versus* tamoxifen as first-line therapy of advanced breast cancer in postmenopausal women: analysis of survival and update of efficacy from the International Letrozole Breast Cancer Group. *J Clin Oncol* 21: 2101–2109

25 Mouridsen H, Gershanovich M, Sun Y, Perez-Carrion R, Boni C, Monnier A, Apffelstaedt J, Smith R, Sleeboom HP, Janicke F et al. (2001) Superior efficacy of letrozole *versus* tamoxifen as first-line therapy for postmenopausal women with advanced breast cancer: results of a phase III study of the International Letrozole Breast Cancer Group. *J Clin Oncol* 19: 2596–2606

26 Paridaens R (2004) Final results of a randomized Phase III trial comparing exemestane with tamoxifen as first-line hormone therapy for postmenopausal women with metastatic breast cancer. http://www.asco.org/ac/1,1003,_12-002511-00_18-0026-00_19-009893,00.asp.

27 ATAC Trialists' Group (2005) Results of the ATAC (Arimidex, Tamoxifen, Alone or in Combination) trial after completion of 5 years' adjuvant treatment for breast cancer. *Lancet* 365: 60–62

28 BIG 1-98 Collaborative Group (2005) BIG 1-98: A prospective randomized double-blind double-dummy phase III study to evaluate letrozole as adjuvant endocrine therapy for postmenopausal women with receptor- positive breast cancer. *Breast* 14 (Suppl 1):S3

29 ATAC Trialists' Group (2002) Anastrozole alone or in combination with tamoxifen *versus* tamoxifen alone for adjuvant treatment of postmenopausal women with early breast cancer: first results of the ATAC randomised trial. *Lancet* 359: 2131–2139

30 ATAC Trialists' Group (2003) Anastrozole alone or in combination with tamoxifen *versus* tamoxifen alone for adjuvant treatment of postmenopausal women with early-stage breast cancer. Results of the ATAC (Arimidex, Tamoxifen, Alone or in Combination) trial efficacy and safety update analyses. *Cancer* 98: 1802–1810

31 Jakesz R (2004) Benefits of switching postmenopausal women with hormone-sensitive early breast cancer to anastrozole after 2 years adjuvant tamoxifen: combined results from 3,224 women enrolled in the ABCSG Trial 8 and the ARNO 95 trial. http://209.196.53.174/2004/

32 Boccardo F, Rubagotti A, Guglielmini P, Amoroso D, Fini A, Paladini G, Mesiti M, Romeo D, Rinaldini M, Scali S et al. (2005) Switching to anastrozole *versus* continued tamoxifen treatment of early breast cancer. Preliminary results of the Italian Tamoxifen Anastrozole (ITA) trial. *J Clin Oncol* 23(22): 5138–5147

33 Coombes RC, Hall E, Gibson LJ, Paridaens R, Jassem J, Delozier T, Jones SE, Alvarez I, Bertelli G, Ortmann O et al. (2004) A randomized trial of exemestane after two to three years of tamoxifen therapy in postmenopausal women with primary breast cancer. *N Engl J Med* 350: 1081–1092

34 Saphner T, Tormey DC, Gray R (1996) Annual hazard rates of recurrence for breast cancer after primary therapy. *J Clin Oncol* 14: 2738–2746
35 Fisher B, Dignam J, Bryant J, Wolmark N (2001) Five *versus* more than five years of tamoxifen for lymph node-negative breast cancer: updated findings from the National Surgical Adjuvant Breast and Bowel Project B-14 randomized trial. *J Natl Cancer Inst* 93: 684–690
36 Goss PE, Ingle JN, Martino S, Robert NJ, Muss HB, Piccart MJ, Castiglione M, Tu D, Shepherd LE, Pritchard KI et al. (2003) A randomized trial of letrozole in postmenopausal women after five years of tamoxifen therapy for early-stage. *N Engl J Med* 349(19): 1793–1802
37 Cataliotti L, Buzdar A, Noguchi S, Bines J (2004) Efficacy of pre-operative Arimidex (anastrozole) compared with tamoxifen (PROACT) as neoadjuvant therapy in postmenopausal women with hormone receptor-positive breast cancer. *Eur J Cancer Suppl* 2: 69, abs 46
38 Smith IE, Dowsett M, Ebbs SR, Dixon JM, Skene A, Blohmer JU, Ashley SE, Francis S, Boeddinghaus I, Walsh G (2005) Neoadjuvant treatment of postmenopausal breast cancer with anastrozole, tamoxifen, or both in combination: the immediate preoperative anastrozole, tamoxifen, or combined with tamoxifen (IMPACT) multicenter double-blind randomized trial. *J Clin Oncol* 23(22): 5108–5116
39 Smith I, Cataliotti L (2004) Anastrozole *versus* tamoxifen as neoadjuvant therapy for oestrogen receptor-positive breast cancer in postmenopausal women: the IMPACT and PROACT trials. *Eur J Cancer Suppl* 2: 69, abs 47
40 Eiermann W, Paepke S, Appfelstaedt J, Llombart-Cussac A, Eremin J, Vinholes J, Mauriac L, Ellis M, Lassus M, Chaudri-Ross HA et al., Letrozole Neo-Adjuvant Breast Cancer Study Group (2001) Preoperative treatment of postmenopausal breast cancer patients with letrozole: a randomized double-blind multicenter study. *Ann Oncol* 12: 1527–1532
41 Cuzick J, Powles T, Veronesi U, Forbes J, Edwards R, Ashley S, Boyle P (2003) Overview of the main outcomes in breast-cancer prevention trials. *Lancet* 361: 296–300
42 Hillner BE, Ingle JN, Chlebowski RT, Gralow J, Yee GC, Janjan NA, Cauley JA, Blumenstein BA, Albain KS, Lipton A, Brown S (2003) American Society of Clinical Oncology 2003 update on the role of bisphosphonates and bone health issues in women with breast cancer. *J Clin Oncol* 21: 4042–4057
43 Howell A (2005) ATAC Trial Update. *Lancet* 365: 1225–1230
44 Goss PE (2004) NCIC CTG MA.17 Final analysis of updated data. A placebo-controlled trial of letrozole following tamoxifen as adjuvant therapy in postmenopausal women with early stage breast cancer. *Proc Am Soc Clin Oncol* 22: 14S

Aromatase Inhibitors
Edited by B.J.A. Furr



# Lessons from the ArKO mouse

Evan R. Simpson, Margaret E. Jones and Colin D. Clyne

*Prince Henry's Institute of Medical Research, Department of Biochemistry and Molecular Biology, Monash University, Melbourne, Australia*

## Aromatase and its gene

Oestrogen biosynthesis is catalyzed by a microsomal member of the cytochrome P450 superfamily, namely aromatase cytochrome P450 (P450 arom, the product of the *CYP19* gene). The P450 gene superfamily is a very large one, containing over 3000 members in some 350 families, of which cytochrome P450 arom is the sole member of family 19 (see http://drnelson.utmem.edu/cytochromeP450.html). This haem protein is responsible for binding of the $C_{19}$ androgenic steroid substrate and catalyzing the series of reactions leading to formation of the phenolic A ring characteristic of oestrogens.

The human *CYP19* gene was cloned some years ago [1–3], when it was shown that the coding region spans nine exons beginning with exon II. Upstream of exon II are a number of alternative first exons that are spliced into the 5'-untranslated region of the transcript in a tissue-specific fashion (Fig. 1). For example, placental transcripts contain at their 5'-end a distal exon, I.1. This is because placental expression is driven by a powerful distal promoter upstream of exon I.1 [4]. Examination of the Human Genome Project data reveals that exon I.1 is 89 kb upstream of exon II [5]. On the other hand, transcripts in ovary and testes contain at their 5'-ends sequence that is immediately upstream of the translational start site. This is because expression of the gene in the gonads utilizes a proximal promoter, promoter II. By contrast, transcripts in cells of mesenchymal origin, such as adipose stromal cells and osteoblasts, contain yet another distal exon (I.4) located 20 kb downstream of exon I.1 [6]. Adipose tissue transcripts also contain promoter II-specific exonic sequence, but these are undetectable in bone [7].

Splicing of these untranslated exons to form the mature transcript occurs at a common 3'-splice junction that is upstream of the translational start site. This means that although transcripts in different tissues have different 5'-termini, the coding region and thus the protein expressed in these various tissue sites is always the same. However, the promoter regions upstream of each of the several untranslated first exons have different cohorts of response elements, and so regulation of aromatase expression in each tissue is different. Thus the gonadal promoter (II) binds the transcription factors cAMP-response-element

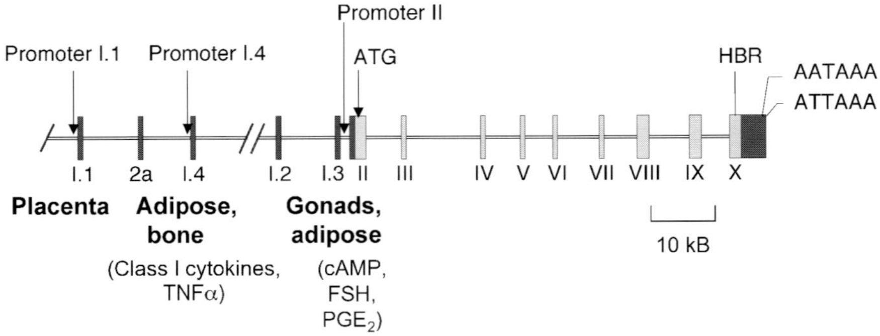


Figure 1. Diagram of the human aromatase (*CYP19*) gene showing tissue-specific promoter usage. The coding region comprises exons II–X. Upstream of the translational start site (ATG) are a number of untranslated exons I which are spliced into the coding region at a common 3'-splice junction in a tissue-specific fashion due to use of the promoters I.1–I.4. The promoters are regulated by the factors indicated. Since this splice junction is upstream of the start of translation, the coding region is always the same, regardless of the tissue of expression. FSH, follicle-stimulating hormone; HBR, haem-binding region; $PGE_2$, prostaglandin $E_2$; TNFα, tumour necrosis factor α.

(CRE)-binding protein (CREB) and steroidogenic factor 1 (SF1), and so aromatase expression in gonads is regulated by cAMP and gonadotrophins. In adipose tissue, promoter II-mediated expression is stimulated by prostaglandin $E_2$ ($PGE_2$). On the other hand, promoter I.4 is regulated by class I cytokines such as interleukin-6, interleukin-11 and oncostatin M, as well as by tumour necrosis factor α. Thus, the regulation of oestrogen biosynthesis in each tissue site of expression is unique (reviewed in [7]) and this leads to a complex physiological situation that makes, for example, interpretation of circulating oestrogen levels as a marker of aromatase activity in specific tissues or in response to specific stimuli very difficult.

## The concept of local oestrogen biosynthesis

Models of oestrogen insufficiency have revealed new and unexpected roles for oestrogens in both males and females. These models include natural mutations in the aromatase gene, as well as mouse knockouts of aromatase and the oestrogen receptors (ERs) [8–13]. In addition, there is one man described with a natural mutation in ERα [14]. Some of the roles of oestrogens apply equally to males and females and do not relate to reproduction; for example the bone, vascular and metabolic syndrome phenotypes.

In postmenopausal women and in men, oestradiol does not appear to function as a circulating hormone, instead it is synthesised in a number of extragonadal sites such as breast, brain and bone where its actions are mainly at the local level as a paracrine or intracrine factor. Thus in postmenopausal women and in men, circulating oestrogens are not the drivers of oestrogen action, rather they reflect the metabolism of oestrogens formed in these extragonadal sites; they are reactive rather than proactive [15]. Importantly, oestrogen biosynthesis in these sites depends on a circulating source of androgenic precursors such as testosterone.

Table 1 shows the plasma steroid levels in postmenopausal women and in men. As can be seen, the levels of oestrone and oestradiol in the plasma of postmenopausal women are extremely low, lower in fact than those in the circulation of men; and, moreover, the levels of circulating testosterone are an order of magnitude greater than those of oestrogens in postmenopausal women. This in itself would suggest that circulating testosterone is better placed to serve as a precursor of oestradiol in target tissues than is circulating oestradiol. On the other hand, the levels of testosterone in the blood of men are an order of magnitude higher than those of women. Significantly, levels of dehydroepiandrosterone (DHEA) and DHEA sulphate (DHEA-S) in the blood of both men and women are orders of magnitude higher than those of the circulating active steroids. In postmenopausal women, the ovaries secrete 25–35% of the circulating testosterone. The remainder is formed peripherally from androstenedione and DHEA produced in the ovaries, and from androstenedione, DHEA and DHEA-S are secreted by the adrenals. However, the secretion of these steroids and their plasma concentrations decrease markedly with advancing age [15, 16].

Figure 2 shows the metabolism of testosterone and oestradiol in a typical target cell [15]. Testosterone in this cell can be derived from the uptake of testosterone or else of androstenedione, DHEA or DHEA-S, all of which can be converted in the target cell to testosterone. Testosterone in turn can act directly on the androgen receptor or else be converted to dihydrotestosterone, which then acts on the androgen receptor. Alternatively, testosterone can be converted to oestradiol that in turn acts on the ER. Testosterone and oestradiol can then leave the cell as such or else be converted to reduced and conjugated metabolites that circulate in the blood at concentrations higher than those of the active steroids [15]. Based on these considerations it is difficult to

Table 1. Plasma steroid concentrations in postmenopausal women and in men

| | Steroid concentration (nM) | |
|---|---|---|
| | Women | Men |
| Testosterone | 0.6 | 12 |
| Androstenedione | 2.5 | 4 |
| Oestrone | 0.10 | 0.13 |
| Oestradiol | 0.04 | 0.10 |
| DHEA | 15 | 10 |
| DHEA-S | 2500 | 2000 |

DHEA, dehydroepiandrosterone; DHEA-S, DHEA sulphate.

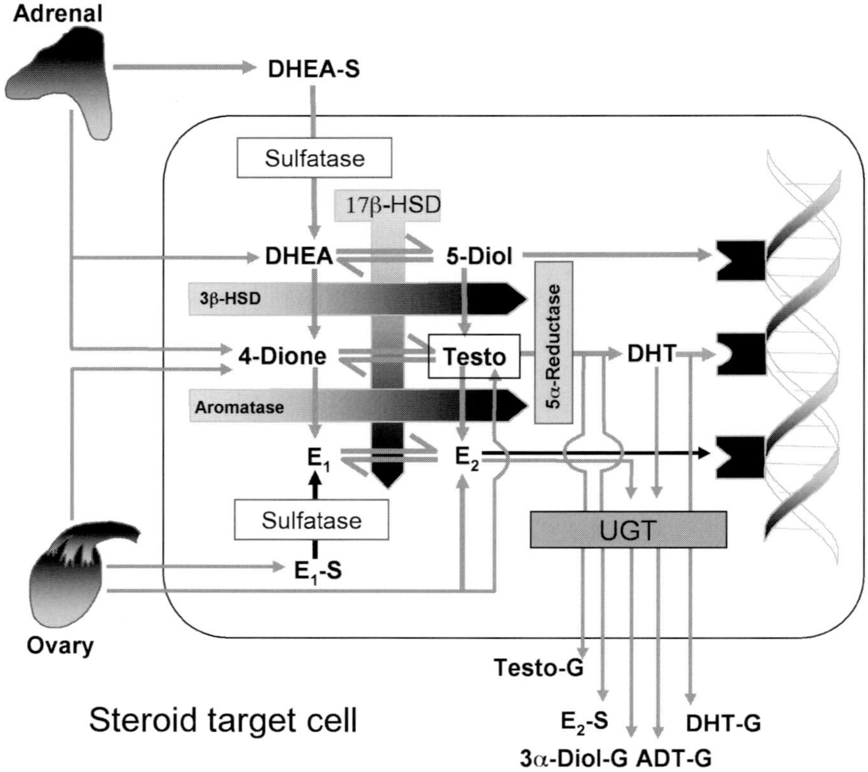


Figure 2. Pathways of metabolism of testosterone and oestradiol in target tissues. Modified from Labrie et al. [16] with permission. DHEA, dehydroepiandrosterone; DHEA-S, DHEA sulphate; HSD, hydroxysteroid dehydrogenase; 5-Diol, 5α/β-androstanediol; 4-Dione, androstenedione; Testo, testosterone; $E_1$, oestrone; $E_2$, oestradiol; DHT, dihydrotestosterone; UGT, UDP-glucuronyl transferase; G, glucuronate; ADT-G, androsterone glucuronide.

see how one can readily equate plasma levels of testosterone and oestradiol to the concentrations that are present in target cells. These considerations lead to the following conclusions regarding the significance of peripheral steroid metabolism: (i) women and men make close to equal amounts of testosterone and oestradiol (say, 50% each rather than 10% in the case of women relative to men) and both have major physiological roles in both sexes; (ii) however, in premenopausal women, most of the testosterone is formed, acts and is metabolized in specific target tissues: it is a paracrine and intracrine factor whereas in men it circulates as a hormone and acts globally; (iii) on the other hand in men most of the oestradiol is formed, acts and is metabolized in specific target tissues whereas in women it circulates as a hormone and acts globally and (iv) finally, in postmenopausal women, in contrast, neither testosterone nor oestradiol function to any extent as a circulating hormone. Both are mainly formed locally in target tissues and act and are metabolized therein.

The power of local oestrogen biosynthesis is illustrated in the case of postmenopausal women with breast cancer [17]. It has been determined that the concentration of oestradiol present in breast tumours of postmenopausal women is at least 20-fold greater than that present in the plasma. With aromatase inhibitor therapy, there is a precipitous drop in the intratumoural concentrations of oestradiol and oestrone together with a corresponding loss of intratumoural aromatase activity, consistent with this activity within the tumour and the surrounding breast adipose tissue being responsible for these high tissue concentrations [18].

In bone, aromatase is expressed primarily in osteoblasts and chondrocytes [19], and aromatase activity in cultured osteoblasts is comparable to that present in adipose stromal cells [20]. Thus it appears that in bone also, local aromatase expression is a major source of oestrogen responsible for the maintenance of mineralization, although this is extremely difficult to prove due to sampling problems. Hence for both breast tumours and for bone, it is likely that circulating oestrogen levels are minimally responsible for the relatively high endogenous tissue oestrogen levels; rather, the circulating levels reflect the sum of local formation in its various sites. This is a fundamental concept for the interpretation of relationships between circulating oestrogen levels in postmenopausal women and oestrogen insufficiency or excess in specific tissues.

The second important point is that oestrogen production in these extragonadal sites is dependent on an external source of $C_{19}$ androgenic precursors, since these extragonadal tissues are incapable of converting cholesterol to the $C_{19}$ steroids [16, 19]. As a consequence, circulating levels of testosterone and androstenedione as well as DHEA and DHEA-S become extremely important in terms of providing adequate substrate for oestrogen biosynthesis in these sites, and therefore differences in the levels of circulating androgens are likely to be important determinants for the maintenance of local oestrogen levels in extragonadal sites.

In this context, it is appropriate to consider why osteoporosis is more common in women than in men and affects women at a younger age in terms of fracture incidence. We have suggested that uninterrupted sufficiency of circulating testosterone in men throughout life supports the local production of oestradiol by aromatization of testosterone in oestrogen-dependent tissues, and thus affords continuing protection against the so-called oestrogen-deficiency diseases. This appears to be important in terms of protecting the bones of men against mineral loss and may also contribute to the maintenance of cognitive function and prevention of Alzheimer's disease [22].

## The aromatase-knockout (ArKO) mouse

In order to investigate the phenotypes resulting from lack of oestrogen, and thereby to understand broader pharmacologically-related side effects of aromatase inhibitors, some years ago we and others generated the aromatase-knockout or ArKO mouse [12, 13, 23, 24]. This was done in our case by replacing most of exon 9 with the neomycin-resistance cassette. Since exon 9 contains many of the amino acids involved in substrate binding, and many of the natural point mutations that result in a complete loss of aromatase activity are located in exon 9, deletion of this exon results in a complete abrogation of aromatase activity. The main features of the phenotype of the ArKO mouse can be summarized as follows: infertility and lack of sexual behaviour in both males and females, progressive defects in folliculogenesis and spermatogenesis; elevated gonadotrophins and testosterone levels; loss of bone mass in both sexes; and a metabolic syndrome with insulin resistance, truncal obesity and hepatic steatosis. Many, but not all aspects of this phenotype are also present in the ERα-knockout and ERα/β-knockout mice (reviewed in [25]). The requirements of oestrogen for male sexual behaviour and for maintenance of male bone mineralization were quite unexpected at the time, but space does not permit discussion of these aspects, which can be found in [26–28]. Instead, we will focus here on the role of oestrogen in energy homeostasis.

## The ArKO mouse and the metabolic syndrome

From the age of 12–14 weeks onwards, ArKO mice develop a progressive phenotype of truncal obesity with increased adiposity in the gonadal and visceral fat pads [13]. Magnetic resonance imaging (MRI) data show that ArKO females have three or four times as much adipose as wild-type females, whereas males have twice as much, so this phenotype of increased adiposity is more marked in the females than in the males. As might be expected then, serum leptin levels are also elevated, as shown in Table 2, so that by 1 year of age, ArKO females have three times as much circulating leptin as do the wild-type

Table 2. Serum leptin levels in ArKO and wild-type mice

| | Serum leptin level (ng/ml) | |
|---|---|---|
| | Female | Male |
| 4 months | | |
| ArKO | 8.18 ± 0.78 (5)* | 8.79 ± 1.83 (6)* |
| Wild-type | 2.92 ± 0.68 (5) | 3.81 ± 1.00 (7) |
| 1 year | | |
| ArKO | 19.86 ± 4.90 (6)* | 8.47 ± 1.85 (7)* |
| Wild-type | 6.19 ± 2.33 (4)† | 4.89 ± 0.72 (8) |

* At least $P < 0.05$ compared to age-matched wild-type mice.
† At least $P < 0.05$ compared to 4-month old genotype- and sex-matched mice.
Figures in parentheses are numbers of mice. Means ± S.E.M. are shown.

females, whereas males have twice as much, consistent with the degree of adiposity in the males and females.

Measurement of serum insulin reveals that the ArKO mice develop hyperinsulinaemia so that by 1 year of age male ArKO mice have three times the level of circulating insulin as do the wild-types (Tab. 3) [13]. However, serum glucose levels remain steady, indicating that at 1 year of age the animals have not progressed to full type 2 diabetes. In spite of the marked increase in adiposity, there was not such a dramatic increase in body weight, leading us to suspect there could be a decrease in lean body mass. This was found to be the case, suggesting a decrease in skeletal muscle mass [13]. To investigate this, energy-balance studies were conducted. These indicated that there was no change in resting energy expenditure or fat oxidation but there was about a 50% reduction in the glucose oxidation rate. There was also a decrease of about 50% in daily ambulatory movements. Since most glucose oxidation is accounted for by skeletal muscle activity, these results are consistent with the insulin resistance being primarily a function of impaired skeletal muscle activity [13].

We then went on to conduct oestrogen replacement studies by the use of silicone implants containing oestradiol which give plasma levels of oestradiol of around 50 pg/ml, in other words approximately the levels seen at the peak of the oestrous cycle, thus within the physiological range [29]. To our surprise, after 21 days there was a dramatic decrease in the visceral fat masses to levels well below those seen with the wild-type placebo controls. This was largely a function of changes in the volume of the adipocytes since there was little change in adipocyte number. We also examined the levels of enzymes and factors involved in *de novo* fatty acid synthesis such as peroxisome proliferator-activated receptor γ (PPARγ), PPARγ coactivator 1-α (PGC1-α), fatty acid synthase and acetyl-CoA carboxylase, but there were no significant changes in expression of these factors. Instead, the increase in adiposity appeared to be primarily due to an increase in the expression of lipoprotein lipase, the enzyme responsible for hydrolysing triglycerides in chylomicra, micra and very-low-

Table 3. ArKO mice develop insulin resistance

| | Insulin (mU/ml) | Glucose (mM) |
|---|---|---|
| ArKO | | |
| 4 months old | 5.98 ± 1.00 (3) | ND |
| 1 year old | 38.67 ± 11.18 (5)* | 8.52 ± 1.56 (3) |
| Wild-type | | |
| 4 months old | 5.26 ± 0.75 (4) | ND |
| 1 year old | 13.82 ± 3.82 (4) | 8.61 ± 2.02 (3) |

* At least $P < 0.05$ compared to age-matched wild-type mice.
Figures in parentheses are numbers of mice. Means ± S.E.M. are shown.

density lipoprotein such that the resulting free fatty acids and *sn*-2 monoglycerides are taken up by the adipose cells and resynthesized into triglycerides. Expression of this enzyme was elevated 3–4-fold in the ArKO mice [29] and profoundly inhibited by oestradiol replacement.

While conducting these experiments we noticed that the livers of the male ArKO mice were paler in colour than those of the wild-type males or of the females. Microscopic examination revealed that the livers of the male ArKO mice were engorged with lipid, whereas those of the females were not [30] (Fig. 3). Analysis of the lipid content revealed that this was primarily due to a 4–5-fold increase in the triglyceride content of the male ArKO livers. Treatment with oestradiol for 6 weeks effectively blocked this increase in hepatic lipid accumulation. Thus the phenotype of the ArKO mice is characterized by a markedly sexually dimorphic lipid partitioning with the increase in lipid in the case of the females occurring primarily in the visceral adipose depots, whereas in the males there is a shift in lipid deposition such that an increased proportion is deposited in the liver, resulting in marked hepatic steatosis. We also examined the expression of enzymes involved in fatty acid synthesis in the livers of these mice and found that in the males there was a

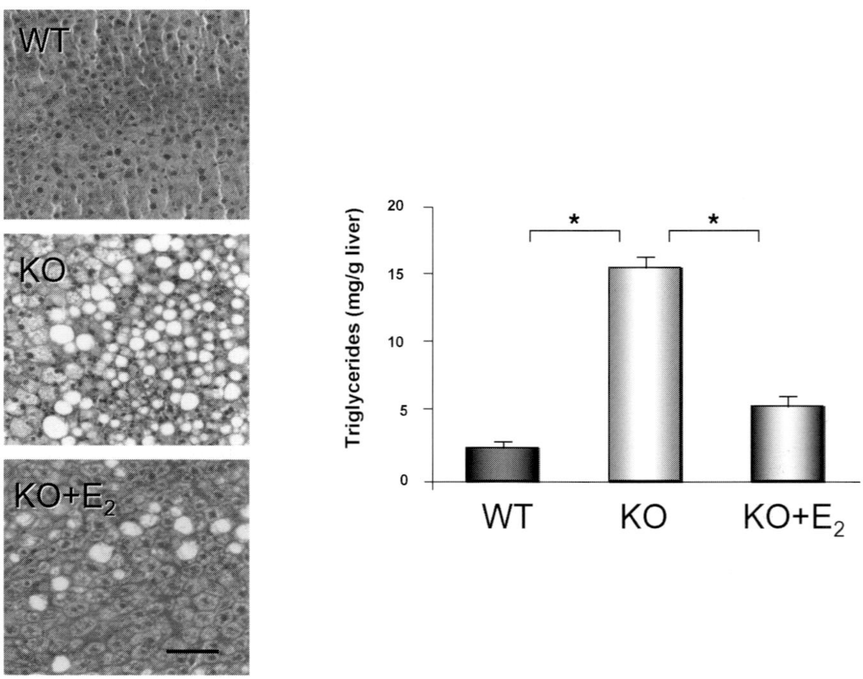


Figure 3. Hepatic phenotype of the male ArKO mouse and the effect of oestradiol replacement. The photomicrographs are representative sections of livers from wild-type (WT) and ArKO (KO) male mice and ArKO mice treated with oestradiol (KO + $E_2$). The histogram on the right shows the corresponding hepatic triglyceride levels. Scale bar: 100 μm)

3–4-fold increase in the expression of fatty acid synthase and of acetyl-CoA carboxylase-α. There was a similar increase in the levels of adipose differentiation related protein (ADRP), a fatty acid transporter. Again, these increases were normalized by oestradiol replacement [30].

In order to understand the basis for this sexually dimorphic phenotype, we are currently examining the hypothalami of the brains of these animals. Previous studies from Gustafsson's laboratory [31] and also the laboratories of Korach and Negishi [32] have indicated that there is a sexually dimorphic pattern of secretion of growth hormone and that this is responsible for the sexually dimorphic imprinting of expression of hepatic P450 enzymes involved in drug and steroid metabolism. For this reason, we examined the arcuate nucleus of these animals, since this is the site of growth hormone-releasing hormone secretion, which is a primary regulator of growth hormone secretion. The arcuate nucleus is also a major site of regulation of feeding behaviour and energy homeostasis. Moreover, pro-opiomelanocortin and neuropeptide Y neurons in the arcuate nucleus are the principal sites of leptin receptor expression and are the source of potent neuropeptide modulators such as melanocortin and neuropeptide Y. TUNEL (terminal deoxynucleotidyl transferase-mediated dUTP nick-end labelling) staining and staining with active caspase 3 revealed a marked increase in apoptosis of tyrosine hydroxylase expressing neurons in the arcuate nucleus of male ArKO but not female ArKO brains. This resulted in a marked loss of tyrosine hydroxylase-positive neurons in the male ArKO arcuate nucleus which is not present in the female [33]. Thus there is a sexually dimorphic loss of dopaminergic neurons in the arcuate nucleus of male ArKO mice. Whether there is a causal relationship between this defect and the sexually dimorphic pattern of lipid accumulation in the ArKO livers remains to be ascertained.

## The metabolic syndrome in humans with natural mutations in aromatase

Currently about a dozen or so individuals have been characterized with natural aromatase mutations, of whom five are men [34–38]. The women so far described have been diagnosed at the time of puberty and placed on oestrogen replacement, so it has not been possible to study their lipid and carbohydrate profiles. Consequently, these studies have been confined to men with aromatase mutations. The most recent study is of an Argentinian male whose phenotype was characterized by Dr Laura Maffei and her colleagues in Buenos Aires and Dr Cesare Carani and his colleagues in Modena, Italy [38]. His metabolic parameters are presented in Table 4. As can be seen, his glucose and insulin levels are markedly elevated and these levels are decreased after oestradiol replacement. He also has acanthosis nigricans. Based on this he was diagnosed as having type 2 diabetes at the age of 29 years. Oestradiol replacement also caused a decrease in total circulating total and low-density lipoprotein

Table 4. Metabolic and liver function parameters of an aromatase-deficient man

| | Before oestradiol treatment | After oestradiol treatment |
|---|---|---|
| Metabolic parameters | | |
| Total cholesterol (mg/dl) | 177 | 110 |
| LDL cholesterol (mg/dl) | 107 | 66 |
| HDL cholesterol (mg/dl) | 31 | 41 |
| Triglycerides (mg/dl) | 199 | 106 |
| Glucose (70–110 μg/dl) | 180 | 144 |
| Insulin (5–30 mU/ml) | 94 | 53 |
| Fructosamine (mM) | 406 | 315 |
| Liver function parameters | | |
| GPT (<37 U/l) | 195 | 70 |
| GOT (<40 U/l) | 108 | 45 |
| γ-GT (<11–50 U/l) | 153 | 42 |

HDL, high-density lipoprotein; LDL, low-density lipoprotein. GPT, glutamic pyruvic transaminase; GOT, glutamic oxaloacetic transaminase; GT, γ-glutamyl transferase.

cholesterol and an increase in high-density lipoprotein cholesterol. His liver-function parameters were also profoundly elevated, as indicated in Table 4, and once again these were markedly reduced upon oestrogen replacement. A liver biopsy revealed substantial macro- and microsteatosis as well as portal vein fibrosis and steatosis. He also had carotid plaques that are unusual in a man of his relative youth and once again these disappeared after oestrogen treatment. Thus this man, with a natural mutation in aromatase, has a metabolic syndrome phenotype that is similar in many ways to that of the male ArKO mice.

## Summary of the metabolic effects of oestrogen

Based on these results, we can conclude that oestrogen has an important role to play in energy homeostasis in both mice and humans. Lack of oestrogen results in the development of a metabolic syndrome. This results in a sexually dimorphic partitioning of lipids such that in males there is profound hepatic steatosis that is not seen in females. Oestrogen administration results in a prompt reversal of these symptoms. We conclude that oestrogen is another hormone synthesized in brain, muscle and adipose tissue that acts to regulate energy homeostasis along with leptin, adiponectin, resistin and cortisol. Because aromatase inhibitors are coming into widespread use as breast cancer therapy and probably also in chemoprevention, potential metabolic disturbances with long-term use of these compounds should be monitored.

## Local aromatase expression in the breast and breast cancer

As indicated previously, aromatase expression in the breast is implicated as the main source of oestrogen driving breast cancer development. Studies to examine aromatase activity and expression in breast cancer quadrants have indicated that this activity is highest in quadrants of the breast containing a tumour [39, 40]. Indeed, there is a gradient of aromatase expression extending from a tumour, such that expression in the tumour-containing quadrant is equal to that in the tumour itself, but double that in a quadrant of the same breast which does not contain tumour, which in turn is double again the expression present in a cancer-free breast [41]. These results suggest that the tumour is elaborating a factor or factors that stimulate aromatase expression within the tumour and in the surrounding adipose tissue.

In order to understand which factor or factors might be responsible, we and others have examined not only total aromatase transcript expression but also expression of promoter-dependent transcripts [41–43] (Fig. 4). In adipose tissue of healthy breast, as indicated above, aromatase expression is low and is driven primarily by a distal promoter I.4, which is regulated by class 1 cytokines and tumour necrosis factor α produced locally in the tissue and acting in a paracrine and autocrine fashion. On the other hand, in the presence of

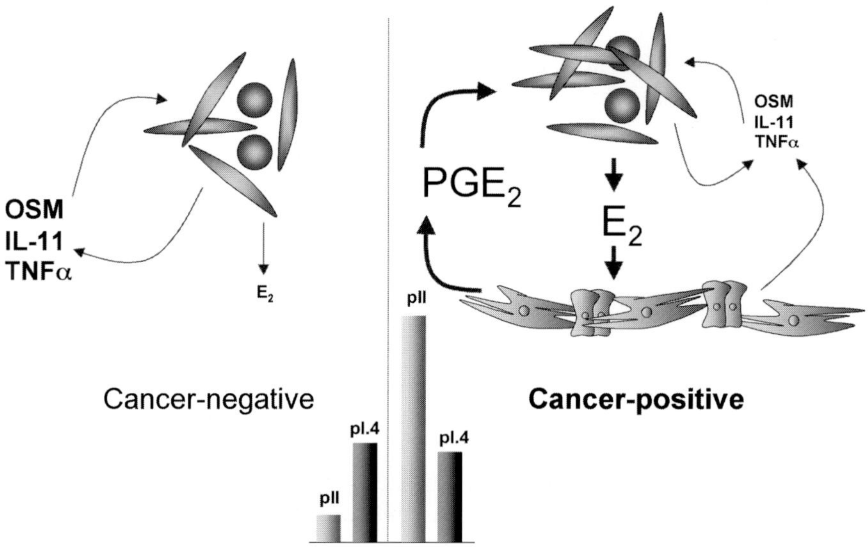


Figure 4. Promoter-specific aromatase transcript expression in cancer-free breast tissue and in proximity to a tumour. The panel on the left shows the situation in healthy breast tissue where promoter (p) I.4 predominates, regulated by cytokines produced by the adipose tissue in a paracrine or autocrine fashion. The panel on the right shows the situation in a tumour-containing breast in which prostaglandin $E_2$ ($PGE_2$) produced by the tumourous epithelium causes switching from promoter I.4 to promoter II and increased aromatase expression. $E_2$, oestradiol; IL-11, interleukin 11; TNFα, tumour necrosis factor α; OSM, oncostatin M.

a tumour, the increase in aromatase expression is due primarily to an increase in expression driven off the proximal gonadal promoter, promoter II. This promoter is regulated by factors that stimulate adenylate cyclase. We reasoned that a likely candidate produced by tumours would be $PGE_2$, and indeed it turned out that $PGE_2$ is a most powerful stimulator of aromatase expression in human breast adipose stromal cells [44, 45] (Fig. 5). Moreover, recent work has indicated that oestrogen has a role itself in upregulating $PGE_2$ synthesis and aromatase in oestrogen receptor-positive breast cancer cells [46]. Moreover, as is well known, cyclo-oxygenase 2 (COX2) is expressed in many breast carcinomas where it correlates with tumour size, high grade and HER2/neu positivity as well as a worse disease-free interval. We would anticipate then that factors that inhibit COX2 activity and thus prostaglandin $E_2$ synthesis would inhibit aromatase expression within the breast. Moreover, since this pathway of regulation of aromatase within the breast is unique, and does not occur within the bone, nor in the ovaries (since the ovaries of postmenopausal women cease to synthesize oestrogens), such inhibition would specifically inhibit oestrogen formation within the breast but would leave other sites of oestrogen formation where it serves an important function – such as bone, brain and the cardiovascular system – protected. Such COX inhibitors are common analgesic drugs, such as aspirin and ibuprofen, so the question arises, are such compounds beneficial in terms of breast cancer therapy?

A number of case-controlled or observational trials have indicated that these compounds are, indeed, of benefit in terms of breast cancer chemoprevention [47–49]. In fact, in one such trial regular use of ibuprofen resulted in as much as a 50% decrease in the incidence of breast cancer over the study period [47]. Several randomized, double-blind, placebo-controlled trials are currently underway to examine the utility of specific COX2 inhibitors as breast cancer therapy [49], although the recent withdrawal of one of them, rofecoxib, as a result of an increased incidence of cardiac events following continuous long-

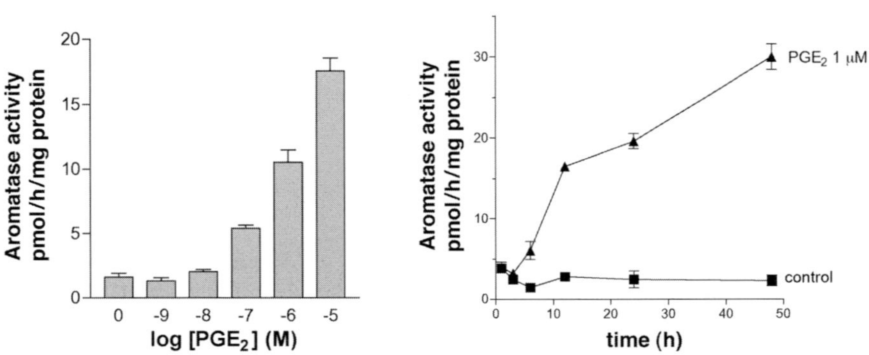


Figure 5. Stimulation of aromatase activity by $PGE_2$ in human breast adipose stromal cells. The left-hand panel shows the dependence on $PGE_2$ concentration whereas that on the right shows a time-course.

term treatment of colon cancer patients [50] may slow or prevent progress in this area. In the meantime, third-generation aromatase inhibitors are proving superior to tamoxifen as first-line adjuvant therapy and neoadjuvant therapy for breast cancer. Moreover, they show benefit as second-line therapy and a dramatic decrease in the incidence of contralateral breast cancer (reviewed in [51]) compared to tamoxifen. They also showed decreased ischaemic cerebral vascular and thromboembolic events as well as decreased endometrial cancer.

However, there are downsides to the use of these compounds. This stems from the fact that since these are highly specific and high-affinity inhibitors of the catalytic activity of aromatase, they inhibit aromatase activity in every site of expression, not only in breast but also in bone, brain and other sites. Not surprisingly, therefore, their use is associated with an increase in bone loss and fracture risk. Interestingly, there is also an increase in arthralgia or inflammatory joint pain [51], and based on the studies discussed earlier in this chapter, it might be anticipated that there is a potential for a poorer lipid profile as well as perhaps development of a metabolic syndrome with long-term use, although as yet there is no evidence for this.

For these reasons, therefore, there will clearly be a benefit if one could specifically inhibit aromatase in the breast but leave other sites of expression such as bone protected. The only way to do this is to inhibit specifically aromatase expression within the breast. The fact that there is a unique pathway of aromatase expression within the breast due to the promoter switching described previously allows, in principle, for this possibility. This leads to the concept of selective aromatase modulators, or SAMs [28], which are to oestrogen synthesis what selective ER modulators (SERMs) are to oestrogen action, and their tissue site specificity is based on the following: (i) the role of oestradiol is as a paracrine and intracrine factor in postmenopausal women and in men; (ii) the tissue-specific regulation of the aromatase gene is based on the use of tissue-specific promoters and (iii) these promoters employ different stimulatory and inhibitory factors in the various tissue-specific sites of expression. Thus, inhibitors of COX2 could serve as the first generation of such SAMs. However, these compounds inhibit the COX enzymes in a ubiquitous fashion and it would clearly be of benefit to specifically inhibit the pathway of aromatase expression within the breast.

## Role of liver receptor homologue-1 (LRH-1) in aromatase expression in the breast

Aromatase expression from promoter II in the ovary requires the presence of activated CREB which binds to CRE in the promoter II sequence [52]. In the ovary, CREB is activated by the signalling pathway that is initiated when follicle-stimulating hormone binds to its receptor and activates adenylate cyclase. In addition to CREB binding to its CRE, activation of the promoter requires the presence of a monomeric orphan member of the nuclear receptor family to bind to a nuclear receptor half-site downstream of the CRE. In the ovary, this factor is SF1. In the case of adipose tissue, no SF1 is present [53], so although $PGE_2$ can substitute for follicle-stimulating hormone in terms of the cAMP signalling pathway, the question arises as to what factor occupies the nuclear receptor half-site to activate promoter 2 in breast adipose tissue. We tested a number of monomeric orphan nuclear receptors known to bind to such a half-site including estrogen-related receptor α (ERRα), Nurr1, Nor1, nerve growth factor-inducible B (NGF1B) and LRH-1 [53]. The only factor that is able to substitute for SF1 in terms of promoter II activation is LRH-1. SF1 and LRH-1 share a high degree of homology and both belong to the NR5A subfamily of nuclear receptors. In contrast to SF1, LRH-1 is expressed in human adipose tissue as well as in human breast tumours, whereas SF1 is not. Using real-time PCR it was found that in adipose tissue LRH-1 is expressed in the mesenchymal preadipocytes rather than in the adipocytes themselves, a similar distribution to that of aromatase. Moreover, upon differentiation of human preadipocytes to the lipid-laden phenotype, LRH-1 expression drops precipitiously, preceding the loss of aromatase expression, suggesting that aromatase expression is dependent on LRH-1. LRH-1 and cAMP activate promoter II synergistically in 3 T3L1 preadipocytes and mutation of the nuclear receptor half-site completely abrogates this action of LRH-1 [53].

Based on these studies, therefore, we can conclude that LRH-1 substitutes for SF1 in human breast preadipocytes to activate aromatase promoter II expression (Fig. 6). Thus, inhibition of LRH-1 would result in loss of aromatase activity in the breast and hence of oestrogen biosynthesis. Therefore, LRH-1 is a potential target for new breast-specific breast cancer therapies,

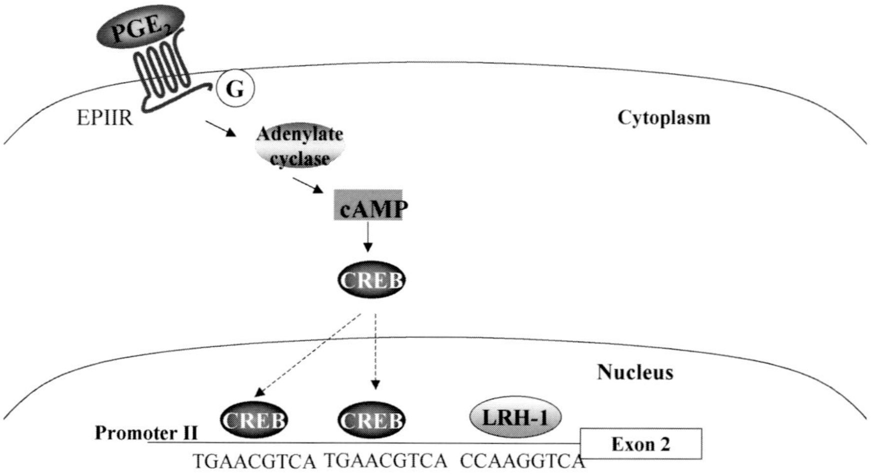


Figure 6. Role of LRH-1 in activation of aromatase promoter II expression in human breast adipose stromal cells. EPIIR, the isoform of the $PGE_2$ receptor which activates adenylate cyclase; TGA(A)CGTCA, the CRE; (CCA)AGGTCA, the nuclear receptor half-site binding element.

since inhibitors of LRH-1 would specifically inhibit aromatase in breast and thus spare oestrogen formation in other tissues. Thus they would serve as SAMS and so could find utility as the next generation of breast cancer therapeutic agents.

*Acknowledgements*
The work from this laboratory described in this chapter was supported by USPHS grant R37AG08174 and by the Victorian Breast Cancer Consortium.

## References

1 Means GD, Mahendroo M, Corbin CJ, Mathis JM, Powell FE, Mendelson CR, Simpson ER (1989) Structural analysis of the gene encoding human aromatase cytochrome P-450, the enzyme responsible for estrogen biosynthesis. *J Biol Chem* 264: 19385–19391
2 Harada N, Yamada K, Saito K, Kibe N, Dohmae S, Takagi Y (1990) Structural characterization of the human estrogen synthetase (aromatase gene). *Biochem Biophys Res Commun* 166: 365–372
3 Toda K, Terashima M, Kamamoto T, Sumimoto H, Yamamoto Y, Sagara Y, Ikeda H, Shizuta Y (1990) Structural and functional characterization of human aromatase. *Eur J Biochem* 193: 559–565
4 Means GD, Kilgore MW, Mahendroo MS, Mendelson CR, Simpson ER (1991) Tissue-specific promoters regulate aromatase cytochrome P450 gene expression in human ovary and fetal tissues. *Mol Endocrinol* 5: 2005–2013
5 Sebastian S, Bulun SE (2001) A highly complex organization of the regulatory region of the human CYP19 (aromatase) gene revealed by the human genome project. *J Clin Endocrinol Metab* 86: 4600–4602
6 Mahendroo MS, Mendelson CR, Simpson ER (1993) Tissue-specific and hormonally controlled alternative promoters regulate aromatase cytochrome P450 gene expression in human adipose tissue. *J Biol Chem* 268: 19463–19470
7 Simpson ER, Zhao Y, Agarwal VR, Michael MD, Bulun SE, Hinshelwood MM, Graham-Lorence S, Sun T, Fisher CR, Qin K, Mendelson CR (1997) Aromatase expression in health and disease. *Rec Prog Horm Res* 52: 185–214
8 Lubahn DB, Moyer JS, Golding TS, Couse JF, Korach KS, Smithies O (1993) Alteration of reproductive function but not prenatal sexual development after insertional disruption of the mouse estrogen receptor gene. *Proc Natl Acad Sci USA* 90: 11162–11166
9 Krege JH, Hodgin JB, Couse JF, Enmark E, Warner M, Mahler JF, Sar M, Korach KS, Gustafsson JA, Smithies O (1998) Generation and reproductive phenotypes of mice lacking estrogen receptor-β. *Proc Natl Acad Sci USA* 95: 15677–15682
10 Couse JF, Hewitt SC, Bunch DO, Sar M, Walker VR, Davis BJ, Korach KS (1999) Postnatal sex reversal of the ovaries in mice lacking estrogen receptors α and β. *Science* 286: 2328–2331
11 Dupont S, Krust A, Gansmuller A, Dierich A, Chambon P, Mark M (2000) Effect of single and compound knockouts of estrogen receptor alpha (ERalpha) and beta (ERbeta) on mouse reproductive phenotypes. *Development* 127: 4277–4291
12 Fisher CR, Graves KH, Parlow AF, Simpson ER (1998) Characterization of mice deficient in aromatase (ArKO) because of targeted disruption of the *cyp19* gene. *Proc Natl Acad Sci USA* 95: 6965–6970
13 Jones MEE, Thorburn AW, Britt KL, Hewitt KN, Wreford NG, Proietto J, Oz OK, Leury BJ, Robertson KM, Yao S, Simpson ER (2000) Aromatase-deficient (ArKO) mice have a phenotype of increased adiposity. *Proc Natl Acad Sci USA* 97: 12735–12740
14 Smith EP, Boyd J, Frank GR, Takahashi H, Cohen RM, Specker B, Williams TC, Lubahn DB, Korach KS (1994) Estrogen resistance caused by a mutation in the estrogen receptor gene in a man. *N Engl J Med* 331: 1056–1061
15 Simpson ER, Rubin G, Clyne C, Robertson K, O'Donnell L, Jones M, Davis S (2000) The role of local estrogen biosynthesis in males and females. *Trends Endocrinol* 5: 184–188
16 Labrie F, Luu-The V, Labrie C, Belanger A, Simard J, Lin SX, Pellitier G (2003) Endocrine and

intracrine sources of androgens in women: inhibition of breast cancer and other roles of androgens and their precursor dehydroepiandrosterone. *Endocrine Rev* 24: 152–182
17 Labrie F, Belanger A, Cusan L, Gomez JL, Candas B (1997) Marked decline in serum concentrations of adrenal C19 sex steroid precursor. *J Clin Endocrinol Metab* 82: 2396–2402
18 Pasqualini JR, Chetrite G, Blacker C, Feinstein MC, Delalonde L, Talbi M, Maloche C (1996) Concentrations of estrone, estradiol and estrone sulfate and evaluation of sulfatase and aromatase activities in pre- and postmenopausal breast cancer patients. *J Clin Endocrinol Metab* 81: 1460–1464
19 DeJong PC, ven de VenJ, Nortier HW, Maitimu-Sneede I, Danker TH, Thijssen JK, Slee PH, Blankenstein RA (1997) Inhibition of breast cancer tissue aromatase activity and estrogen concentratons by the third-generation aromatase inhibitor vorozole. *Cancer Res* 57: 2109–2111
20 Oz OK, Millsaps R, Welch R, Birch J, Zerwekh JE (2001) Expression of aromatase in the human growth plate. *J Mol Endocrinol* 27: 249–253
21 Shozu M, Simpson ER (1998) Aromatase expression of human osteoblast-like cells. *Mol Cell Endocrinol* 139: 117–129
22 Labrie F, Belanger A, Luu-The V, Labrie C, Simond J, Cusan L, Gomez JL, Candas B (1998) DHEA and the intracrine formation of androgens and estrogens in peripheral target tissues: its role during aging. *Steroids* 63: 322–328
23 Honda S, Harada N, Takagi Y, Maeda S (1998) Disruption of sexual behaviour in male aromatase-deficient mice lacking exons 1 and 2 of the *cyp19* gene. *Biochem Biophys Res Commun* 252: 445–449
24 Nemoto Y, Toda K, Ono M, Fujikawa-Adachi K, Saibara T, Onishi S, Enzan H, Okada T, Shizuta Y (2000) Altered expression of fatty acid metabolizing enzymes in aromatase-deficient mice. *J Clin Invest* 105: 1819–1825
25 Couse JF, Korach KS (1999) Estrogen receptor null mice: what have we learned and where will they lead us? *Endocrine Rev* 20: 358–417
26 Ogawa S, Chester AE, Hewitt SC, Walker VR, Gustafsson JA, Smithies O, Korach KS, Pfaff DW (2000) Abolition of male sexual behaviours in mice lacking estrogen receptors alpha and beta. *Proc Natl Acad Sci USA* 97: 14737–14741
27 Oz OK, Zerwekh JE, Fisher G, Graves K, Nann L, Millsaps R, Simpson ER (2000) Bone has a sexually dimorphic response to aromatase deficiency. *J Bone Mineral Res* 15: 507–514
28 Simpson ER, Clyne CD, Rubin G, Boon WC, Robertson K, Britt K, Speed C, Jones ME (2002) Aromatase – a brief review. *Annu Rev Physiol* 64: 93–127
29 Misso M, Murata Y, Boon W-C, Jones ME, Britt KL, Simpson ER (2003) Cellular and molecular characterization of the adipose phenotype of the aromatase-deficient mouse. *Endocrinology* 144: 1474–1480
30 Hewitt KN, Pratis K, Jones ME, Simpson ER (2004) Estrogen replacement reverses the hepatic steatosis phenotype in the male aromatase knockout (ArKO) mouse. *Endocrinology* 145: 1842–1848
31 Morgan ET, MacGeoch C, Gustafsson J-A (1985) Hormonal and developmental regulation of expression of the hepatic microsomal steroid 16α-hydroxylase cytochrome P450 apoprotein in the rat. *J Biol Chem* 260: 11895–11898
32 Sueyoshi T, Yokomori N, Korach KS, Negishi M (1999) Developmental action of estrogen receptor-α feminizes the growth hormone-stat 5b pathway and expression of Cyp2a4 and Cyp2d9 genes in mouse liver. *Mol Pharmacol* 561: 473–477
33 Hill RA, Pompolo S, Jones ME, Simpson ER, Boon WC (2004) Estrogen deficiency leads to apoptosis in dopaminergic neurons in the medial preoptic area and arcuate nucleus of male mice. *Mol Cell Neurosci* 27: 466–476
34 Morishima A, Grumbach MM, Simpson ER, Fisher C, Qin K (1995) Aromatase deficiency in male and female siblings caused by a novel mutation and the physiological role of estrogens. *J Clin Endocrinol Metab* 80: 3689–3698
35 Carani C, Qin K, Simoni M, Faustini Fustini M, Serpente S, Boyd J, Korach KS, Simpson ER (1997) Effect of testosterone and estradiol in a man with aromatase deficiency. *N Engl J Med* 337: 91–95
36 Bilezikian JP, Morishima A, Bell J, Grumbach MM (1998) Increased bone mass as a result of estrogen therapy in a man with aromatase deficiency. *N Engl J Med* 339: 599–603
37 Hermann BL, Saller B, Janssen OE, Gocke P, Bockish A, Sperling H, Mann K, Broecker M (2002) Impact of estrogen replacement therapy in a male with congenital aromatase deficiency caused by

a novel mutation in the CYP19 gene. *J Clin Endocrinol Metab* 87: 5476–5484
38 Maffei L, Murata Y, Rochira V, Tubert G, Aranda C, Vazquez M, Clyne CD, Davis S, Simpson ER, Carani C (2004) Dysmetabolic syndrome in a man with a novel mutation of the aromatase gene: effects of testosterone, alendronate and estradiol treatment. *J Clin Endocrinol Metab* 89: 61–70
39 O'Neill JS, Elton RA, Miller WR (1988) Aromatase activity in adipose tissue from breast quadrants: a link with sumor site. *Br Med J* 296: 741–743
40 Bulun SE, Price TM, Aitken J, Mahendroo MS, Simpson ER (1993) A link between breast cancer and local estrogen biosynthesis suggested by quantification of breast adipose tissue aromatase P450 transcripts using competitive polymerase chain reaction after reverse transcription. *J Clin Endocrinol Metab* 77: 1622–1628
41 Agarwal VR, Bulun SE, Leitch M, Rohrich R, Simpson ER (1996) Use of alternative promoters to express the aromatase cytochrome P450 (CYP19) gene in breast adipose tissues of cancer-free and breast cancer patients. *J Clin Endocrinol Metab* 81: 3843–3849
42 Harada N, Utsume T, Takagi Y (1993) Tissue-specific expression of the human aromatase cytochrome P450 gene by alternative use of multiple exons 1 and promoters, and switching of tissue-specific exons 1 in carcinogenesis. *Proc Natl Acad Sci USA* 90: 11312–11316
43 Zhou C, Zhou D, Esteban J, Murai J, Siiteri PK, Wilczynski S, Chen S (1996) Aromatase gene expression and its exon I usage in human breast tumours. Detection of aromatase messenger RNA by reverse transcription-polymerase chain reaction. *J Steroid Biochem Mol Biol* 59: 163–171
44 Zhao Y, Agarwal VR, Mendelson CR, Simpson ER (1996) Estrogen biosynthesis proximal to a breast tumor is stimulated by PGE2 via cyclic AMP, leading to activation of promoter II of the CYP19 (aromatase) gene. *Endocrinology* 137: 5739–5742
45 Richards JA, Brueggemeier RW (2003) Prostaglandin E2 regulates aromatase activity and expression in human adipose stromal cells via two distinct receptor subtypes. *J Clin Endocrinol Metab* 88: 2810–2816
46 Frasor J, Danes JM, Komm B, Chang K, Lyttle CR, Katzenellenbogen BS (2003) Profiling of estrogen up-and down-regulated gene expression in human breast cancer cells: Insights into gene networks and pathways underlying estrogenic control of proliferation and cell phenotype. *Endocrinology* 144: 4562–4574
47 Harris RE, Chlebowski RT, Jackson RD, Frid DJ, Ascenseo JL, Anderson G, Loar A, Rodabough RJ, White E, McTiernan A, Women's Health Initiative (2003) Breast cancer and non-steroidal anti-inflammatory drugs: prospective results from the Women's Health Initiative. *Cancer Res* 63: 6096–6101
48 Terry MB, Gammon MD, Zhang FF, Tawfik H, Teitelbaum SL, Britton JA, Subboramaiah K, Dannenberg AJ, Neugut AL (2004) Association of frequency and duration of aspirin use and hormone receptor status with breast cancer risk. *JAMA* 291: 2433–2440
49 Arun B, Goss P (2004) The role of COX-2 inhibition in breast cancer treatment and prevention. *Semin Oncol* 31 (2 suppl. 7): 22–29
50 Editorial Lancet (2004) Vioxx: an unequal partnership between safety and efficacy. *Lancet* 364: 1287–1288
51 Howell A, Dowsett M (2004) Endocrinology and hormone therapy in breast cancer: aromatase inhibitors *versus* antiestrogens. *Breast Cancer Res* 6: 269–274
52 Michael MD, Kilgore MW, Morokashi K, Simpson ER (1995) Ad4BB/SF1 regulates cyclic AMP-induced transcription from the proximal promoter (PII) of the human aromatase P450 (CYP19) gene in the ovary. *J Biol Chem* 270: 13561–13566
53 Clyne CD, Speed CJ, Zhou J, Simpson ER (2002) Liver receptor homologue-1 (LRH-1) regulates expression of aromatase in preadipocytes. *J Biol Chem* 277: 20591–20597

Aromatase Inhibitors
Edited by B.J.A. Furr



# Possible additional therapeutic uses of aromatase inhibitors

Barrington J.A. Furr

*Global Discovery, AstraZeneca, Mereside, Alderley Park, Macclesfield, Cheshire SK10 4TG, UK*

## Introduction

Several excellent chapters in this book describe the clinical utility of aromatase inhibitors in the treatment of breast cancer. It is true to say that use of third-generation aromatase inhibitors has had a major therapeutic impact: emerging clinical evidence for some of them shows that they can achieve superior efficacy to tamoxifen, the gold standard of endocrine care for more than two decades.

In contrast to the intensive research on use of aromatase inhibitors in breast cancer and a plethora of publications on this topic, there have been few studies on applications to other diseases where oestrogen contributes to induction, maintenance or progression of the disease state. There is extensive evidence that oestrogen withdrawal with tamoxifen has shown benefit in a range of diseases first reviewed comprehensively by Furr and Jordan [1]. A list of diseases where tamoxifen has been investigated are shown in Table 1. Clear evidence of activity is seen in endometrial cancer, male and female infertility, benign breast disease, delayed puberty, suppression of lactation, gynaecomastia and menometorrhagia. Minor effects have been observed in mostly small trials in ovarian, prostate, renal, colorectal and pancreatic cancer and possibly meningioma. More convincing results have been seen in desmoid tumours.

Evaluation of aromatase inhibitors is therefore justified in those diseases where some encouragement has been seen with tamoxifen therapy. It must be emphasized that this relates only to men, postmenopausal women and female patients with inadequate ovarian function. In premenopausal women it will be necessary to abrogate ovarian function with luteinizing hormone-releasing hormone (LHRH) agonists or antagonists if aromatase inhibitors are to be able to exert maximum effects.

This chapter examines the data available currently on the utility of aromatase inhibitors in the range of diseases where oestrogen appears to be at least partly responsible for symptoms or progression.

Table 1. List of diseases where therapeutic utility of the antioestrogen tamoxifen has been investigated

| Disease | Evidence of efficacy |
|---|---|
| Endometrial cancer | Yes, some good responses |
| Prostate cancer | Minor responses only |
| Ovarian carcinoma | Some responses in limited trials |
| Renal carcinoma | Minor responses only |
| Melanoma | No striking activity |
| Colorectal tumours | Minor responses |
| Gastric cancer | No responses |
| Oesophageal cancer | No responses |
| Pancreatic cancer | Few responses |
| Liver cancer | No activity |
| Meningioma | Possible stabilization |
| Pituitary tumours | Yes, in some prolactinomas |
| Desmoid tumours | Some good remissions |
| Female infertility | Yes, good responses in some patients |
| Male infertility | Yes, good responses in some patients |
| Benign breast disease | Yes, good responses |
| Suppression of lactation | Yes |
| Gynaecomastia | Yes |
| Menometorrhagia | Yes |
| Delayed puberty | Yes |

## Malignant disease

### *Ovarian cancer*

A number of papers describe preliminary studies with aromatase inhibitors for the treatment of ovarian cancer. An open-label phase 2 study in women with ovarian cancer described results from 60 patients treated with 2.5 mg of letrozole at the time of relapse, indicated by elevation of the marker, CA-125. Fifty patients were evaluated for response by Union Internationale Contre le Cancer criteria. Responses were modest with no objective responses but 10 (20%) had stable disease confirmed by scan. Five patients in the group did show a reduction of greater than 50% in the biomarker CA-125. Tumours in the stable group had significantly higher oestrogen receptor (ER) content than those in the progressing group. The drug was well tolerated [2].

In another study, 27 patients with relapsed ovarian cancer were treated with 2.5 mg of letrozole [3]. This study was slightly more positive. Of 21 evaluable patients, one showed a complete response by World Health Organization criteria and two partial responses. A fourth patient showed a CA-125 response and five additional patients had stable disease. Again, letrozole was well tolerated. However, in this study there was no correlation between tumour response and ER or progesterone receptor (PR) expression.

Anastrozole has been studied in combination with the epidermal growth factor tyrosine kinase inhibitor, gefitinib, in a range of recurrent asymptomatic Mullerian cancers including ovarian, peritoneal and fallopian tube carcinoma that were ER- and/or PR-positive [4]. Thirty-five women were enrolled of whom 30 had ovarian cancer. Of the 23 women evaluable, one had a complete response and 14 stable disease. Toxicity was tolerable. It is unclear in this study whether anastrozole, gefitinib or the combination was responsible for the responses seen.

The overall conclusion is that aromatase inhibitors have only modest activity in relapsed ovarian cancer but because of their good tolerability could be considered worthwhile in some frail patients who are unsuitable for intensive chemotherapy.

### *Endometrial cancer*

Since the endometrium is an oestrogen-responsive tissue and endometrial cancers express both ER and PR, it was logical to examine the impact of aromatase inhibitors on tumour growth and to compare any activity with progestins, the current mainstay of therapy for this disease. Moreover, aromatase inhibitors have been shown to reduce proliferation and increase apoptosis in endometrial cancer cells *in vitro* [5, 6].

The overall conclusion from these studies on endometrial carcinoma is that aromatase inhibitors have minimal activity in patients with advanced recurrent tumours but there is limited information on impact in earlier disease that has not been influenced by pretreatment with progestins.

Rose et al. [7] described a phase II trial in which 1 mg of anastrozole daily was given to 23 patients with advanced or recurrent endometrial cancer not curable by either surgery or radiation therapy. Only two partial responses were noted and two patients had short-term stable disease. The toxicity was mild except that one case of venous thromboembolism was reported but it was not clear whether this was drug-related. The authors concluded that the poor response may have been due, in part, to inclusion of aggressive serous and clear cell tumours that are regarded as non-hormone-responsive and the fact that most of these tumours were poorly differentiated.

Berstein et al. [8] studied the effect of 2.5 mg of letrozole on 10 postmenopausal women with previously untreated endometrial cancer for 14 days prior to surgery. In two patients, pain relief in the lower abdomen and/or reduction in uterine discharge were reported. In three cases, there was a surprising marked reduction of endometrial volume (mean 31.1%) by ultrasound in such a short duration of treatment. Again the drug was well tolerated but the duration of treatment was too short to draw any clear conclusions about its likely long-term benefits.

Letrozole has also been studied in a multi-centre Canadian phase 2 trial in women with recurrent or advanced endometrial cancer; serous and clear cell tumours were excluded. Thirty-two patients were treated with 2.5 mg of letrozole. There was one complete response and it may be noteworthy that this was in a patient with metastatic disease but who had not received prior hormonal therapy. Two patients showed partial responses and 11 had stable disease with a medium duration 6–7 months. One of the two patients who showed a partial response also had no previous hormone therapy [9].

Toxicity was mild with only one incidence of grade 3 depression and one of venous thrombosis. This trial suggests that aromatase inhibitors do have some activity in patients with endometrial cancer and that patients with earlier disease that have not received prior hormone therapy should be investigated in more detail.

Endometrial stromal sarcoma is a relatively rare type of endometrial cancer that is relatively indolent but is known to be hormone-responsive and to express ER, PR and aromatase [10, 11]. Maluf et al. [12] described the first case of recurrent endometrial stromal sarcoma treated with 2.5 mg of letrozole in a post-menopausal woman who had received prior surgery, radiation, progestins and tamoxifen therapy. The tumour area showed a 67% reduction (partial response) and there was a complete response of nodules in the right anterior abdominal wall and sub-capsular liver implant. Similarly, Leunen et al. [13] showed a first-line hormonal response to 2.5 mg of letrozole. Spano et al. [14] reported complete response in two patients with endometrial stromal sarcoma, metastatic to the lung, following treatment with the first-generation aromatase inhibitor aminoglutethimide (500 mg four times daily). Reich and Regauer [15] have entered 12 women into a trial of post-operative therapy but no results have yet been described.

The overall conclusion is that endometrial stromal sarcoma may be particularly amenable to aromatase inhibitor therapy but that more comprehensive studies need to be undertaken to put any position they may have in therapy in full perspective.

### *Prostate cancer*

Studies of the use of aromatase inhibitors in prostate cancer have been universally disappointing. A phase 2 study of 1 mg of anastrozole daily in men with advanced prostate cancer refractory to medical or surgical orchidectomy stopped after no objective responses were seen in the first 14 patients treated. Minimal improvements in bone pain were reported in two patients and 10 showed a reduction of >50% in PSA but with no impact on tumour dimensions [16].

Similarly, Smith et al. [16] showed no effect of 2.5 mg daily in 43 men with androgen-independent prostate cancer. In this study, only one patient showed a reduction in PSA of greater than 50%. Treatment was well tolerated.

Exemestane may actually stimulate tumour growth as three out of four patients had a significant increase in bone pain only a few days after starting treatment and there was clear PSA progression; both of these were reversed on drug withdrawal [18]. This may be due to some androgenic activity in this steroidal aromatase inhibitor and serves to emphasize that not all third-generation aromatase inhibitors have identical pharmacological effects.

### *Liver cancer*

Therapy for hepatocellular carcinoma is inadequate and often compromised by cirrhosis. There is some evidence that this tumour may be oestrogen-responsive, although tamoxifen has little value. Nevertheless, Grosh et al. [19] studied the effect of 1 mg of anastrozole daily in 14 patients with hepatocellular carcinoma. Four patients were said to have stable disease for up to 24 weeks but no objective responses were seen. The drug was well tolerated. It is concluded that aromatase inhibitors are unlikely to have any real therapeutic value in liver cancer.

## Non-malignant disease

### *Female infertility*

Tamoxifen and clomiphene have been used for several decades for the treatment of anovulatory infertility [1], so it is unsurprising that a number of studies have investigated the role of aromatase inhibitors in infertile women. Mitwally and Casper [20] published the first report of use of letrozole on induction of ovulation in women with polycystic ovaries. Promising results were obtained showing that letrozole had no adverse antioestrogen-like effects on endometrial thickness and cervical mucus, probably because of its relatively short half-life.

Four studies describe the impact of aromatase inhibitors in women with polycystic ovary syndrome (PCOS). Mitwally and Casper [21] administered 2.5 mg of letrozole on days 3–7 of the menstrual cycle of 12 patients who had achieved inadequate responses to clomiphene. Ovulation occurred in nine and pregnancy ensued in three women; there was no compromise of endometrial growth. In a similar study [22], 22 infertile women with PCOS resistant to clomiphene were given 2.5 mg of letrozole daily on days 3–7 of the menstrual cycle. Ovulation occurred in 84.4% of treatment cycles and pregnancy ensued in six patients (27%). Again, endometrial thickness was not affected. In this study, 18 additional patients were given 2 mg of anastrozole daily that appeared to be less effective in inducing ovulation (60% of cycles) and pregnancy (16.6%). In a comparison of treatment with 1 mg of anastrozole and clomiphene in 50 women with anovulatory infertility, there was no difference in ovulation rate, number of dominant follicles and pregnancy rate but endometrial thickness was significantly higher in those treated with anastrozole [23]. In the largest study to date Elnashar [24] described 44 patients with

PCOS resistant to clomiphene treated with letrozole for 5 days. An ovulation rate of 54% and a pregnancy rate of 29% were achieved.

Aromatase inhibitors have also been used with some success in women with ovulatory infertility. Ten infertile women who were ovulatory but had inadequate responses to clomiphene were given 2.5 mg of letrozole on days 3–7 of the menstrual cycle. A mean of 2.3 mature follicles were produced without any impact on endometrial thickness; pregnancy ensued in one patient [21]. Similar results were seen in 19 ovulatory normal volunteers treated with either 2.5 mg of letrozole or 50 mg of clomiphene daily on days 5–9 after menstruation [25]. It was concluded that letrozole caused comparable stimulation of ovarian folliculogenesis to clomiphene but, unlike clomiphene, had no adverse effects on endometrial thickness or pattern at mid-cycle.

Aromatase inhibitors have also been studied in assisted-reproduction programmes both to provide eggs for implantation and in attempts to reduce the amount of expensive follicle-stimulating hormone (FSH) preparations used in the schedules. Sammour et al. [26] compared the effects of clomiphene and letrozole in 49 patients with unexplained infertility undergoing super-ovulation prior to intrauterine insemination. They found that, although clomiphene induced more mature follicles by the time of administration of human chorionic gonadotrophin (hCG), the endometrium was thinner than in the letrozole group. Probably as a consequence, the pregnancy rate was three times higher in the letrozole group. Mousavi-Fatemi et al. [27] confirmed the finding that fewer mature follicles developed in letrozole-treated women than in those given clomiphene. In another study El Helw et al. [28] used a much higher dose of 20 mg of letrozole as a single dose on day 3 of the menstrual cycle in 53 randomized patients and achieved a marginally higher pregnancy rate with letrozole (18.2%) compared with clomiphene (11.5%).

In a number of studies letrozole has been shown to reduce the amount of FSH required to induce super-ovulation in infertile women but the ideal dosing regimen has yet to be established. Mitwally and Casper [29] studied the use of letrozole in 12 infertile women who responded poorly to FSH administration by producing fewer than three follicles. Letrozole was given at a dose of 2.5 mg on days 3–7 of the menstrual cycle. The patients showed an enhanced response to FSH in terms of increased numbers of mature follicles; a pregnancy rate of 21% was achieved and a reduced dose of FSH required. The results have been confirmed and extended in two further reports by these authors [30, 31].

Comparison of a short protocol with a gonadotrophin-relasing hormone (GnRH) agonist, FSH with letrozole and FSH and a GnRH antagonist in patients with poor responses to FSH showed that the FSH dose needed to achieve a satisfactory ovulation rate was lower in the letrozole group and endometrial thickness was improved; the pregnancy rate was 16.7% following letrozole compared with 7.7% in the control group [32].

Healey et al. [33] confirmed the findings that FSH could be spared if patients were given 5 mg of letrozole but that endometrial thickness was compromised, which is at variance with most other studies. In this study, gonadotrophins were administered either alone from day 3 or in combination with 5 mg of letrozole from day 5 of the menstrual cycle. Ovulation was triggered by administration of hCG when the dominant follicle reached 18 mm in diameter. Patients who were co-administered letrozole required fewer gonadotrophin ampoules and developed more follicles with a diameter of greater than 14 mm; pregnancy rate did not differ between the groups and was around 20%. It seems likely that the reason for the adverse impact on endometrial thickness in the letrozole group was due to timing of drug administration. In the study of Healey et al. [33] ovulation was induced about 4 days after stopping a dose of letrozole that was twice the standard dose. Taking the half-life of letrozole into consideration, it seems likely that therapeutically active concentrations of the drug were present at the time of hCG administration that might account for reduced oestrogen production and impaired endometrial thickness [34].

### *Endometriosis*

Endometriosis is known to be an oestrogen-responsive disease and is stimulated by ovarian production of oestrogen in premenopausal women but there may also be a local tissue component as endometriotic tissue expresses aromatase [35–37]. Moreover, 1 mg of anastrozole caused a significant improvement in a postmenopausal woman with recurrent severe endometriosis maintained following oophorectomy. The size of endometriotic lesions was reduced and pelvic pain was relieved [38, 39]. In a similar study in a 31-year-old woman who had undergone ovariectomy for severe endometriosis but in whom the disease recurred, 2.5 mg of letrozole caused significant decreases in both pelvic pain and dyspareunia accompanied by significant decreases in oestrogen [40].

These results imply that in some women with minimal or no ovarian function sufficient oestrogen can be produced either peripherally or within the endometriotic lesions to maintain active disease and that aromatase inhibitors will produce compelling improvements.

A number of reports of use of aromatase inhibitors in premenopausal women with endometriosis have also appeared [41–46]. However, in none of these studies was an aromatase inhibitor used alone, probably because of their inability to reduce sufficiently the high concentrations of circulating oestrogen due to the high ovarian aromatase expression and aromatase substrate (androgen) concentrations. Four reports describe the use of the depot GnRH agonist, Zoladex, given monthly alone and in combination with 1 mg of anastrozole. Over 40 patients were randomized to this treatment. Remission of disease was achieved and restoration of fertility was seen in 10 patients in the combined treatment group; this was significantly more than in patients treated with Zoladex alone where fertility was only restored in four patients [42]. Perhaps the most important observation was that recurrence of the disease was more common following drug withdrawal of Zoladex alone than with the combination after both 6 and 12 months [41–43, 45]. The medium time to symptom recurrence was also significantly longer in the combination group [45].

In two other studies, aromatase inhibitors have been combined with progestins both to attempt to suppress gonodotrophins and oestrogen secretion and to 'antagonize' oestrogen action. Combination of 2.5 mg of letrozole and 2.5 mg of norethindrone acetate for 6 months caused complete remission of peritoneal lesions in 10 women with endometriosis. American Society for Reproductive Medicine scores and pelvic pain decreased significantly during treatment [44]. In a smaller study on two women with endometriosis 1 mg of anastrozole was combined with 200 mg of oral progesterone daily for 21 days of six 28-day cycles. Treatment resulted in rapid progressive reduction in symptoms and maintenance of remission for over 2 years after treatment. Absence of lesions was observed in one patient at follow-up laparoscopy and both patients became pregnant [46].

The conclusion that can be drawn is that in premenopausal women with endometriosis aromatase inhibitors do offer additional benefit to standard treatment with either GnRH agonists or progestins.

### *Fibromatosis*

Since fibroids are also known to be oestrogen-responsive it is surprising that there are few studies on the impact of aromatase inhibitors on this disease either to cause regression or to limit the need for hysterectomy by reducing disease burden and allowing myomectomy. Indeed, the only paper on this topic that was identified described the administration of 2.5 mg of letrozole to a hysterectomized-oophorectomized woman who retained inoperable pelvic fibromatosis [47]. A good response was observed and there was significant reduction in size of some of the pelvic masses by computed tomography scan and complete resolution of others; this was associated with complete freedom from symptoms.

### *Male infertility*

There is clear evidence that anastrozole stimulates the hypothalamus–pituitary–testes axis in rats [48]. This is manifest by significant increases in plasma FSH and testosterone and in testes weight. It is, therefore, logical to evaluate their effects in men with inadequate gonadal function. In a major study involving over 100 infertile men, 1 mg of anastrozole daily for a mean duration for 4.7 months caused a significant increase in serum testosterone and a reduction in serum oestradiol, except in those patients with Klinefelter's syndrome. There was a significant increase in mean semen volume (2.9 *versus* 3.5 ml), sperm concentration (5.5 *versus* 15.6 million sperm/ml) and motility index in 25 oligospermic men but, not unexpectedly, no effect in azoospermic patients [49]. Similar results were found with the first-generation aromatase inhibitor, testolactone, given twice daily at a dose of 50 mg.

In a smaller study in 10 men with idiopathic hypogonadotrophic hypogonadism with premature ejaculation, 2-week therapy with 1 mg of anastrozole daily caused increased serum luteinizing hormone (LH) and testosterone and reduced serum oestradiol. Perhaps not unexpectedly, there was no effect on premature ejaculation [50].

Leder et al. [51] have examined the effect of anastrozole on the depressed levels of testosterone in elderly men. Thirty-seven elderly men (aged 62–74) were randomized to 1 mg of anastrozole given either daily or twice weekly or placebo for 12 weeks. There was a significant increase in serum LH (5.1 to 7.9 units/l), total testosterone (343 to 572 ng/dl) and bioavailable testosterone (99 to 207 ng/dl) in patients given 1 mg of anastrozole daily. Serum oestradiol decreased (26 to 17 pg/ml). These results show that daily administration of 1 mg of anastrozole can increase serum bioavailable and total testosterone in elderly men with mild hypogonadism to the normal youthful range. However, any physiological benefit of these changes remains to be determined.

It can be concluded that aromatase inhibitors do stimulate testis function in men and are worthy of further study to determine whether these changes have an impact on fertility in infertile patients or on sexual function in ageing men.

### *Puberty*

There have been a number of studies in three different situations related to puberty: delayed puberty, precocious puberty and pubertal gynaecomastia.

Delayed puberty causes relatively short spinal height and may also result in reduced skeletal integrity of the spine so predisposing such adolescents to high risk of fracture later in life [52]. Administration of androgens [53] or anabolic steroids [52] provides effective treatments that advance secondary sexual characteristics and the growth spurt but do not improve final height. There is good evidence that epiphyseal closure is oestrogen-dependent. In men with aromatase deficiency due to gene mutation, the epiphyses of long bones were unfused until the men were well into their twenties and so the men continue to grow. Administration of oestrogen caused rapid epiphyseal closure [55, 56]. There is, therefore, a good case for combining androgen and aromatase inhibitors to treat delayed puberty with the objective of increasing adult height and securing improved bone integrity.

There are two reports of a study of the effect of 2.5 mg of letrozole on patients with delayed puberty [57, 58]. Ten boys were untreated and served as a control group; one group of 12 boys received testosterone enanthate (1 mg/kg) every 4 weeks and another group received the androgen plus 2.5 mg of letrozole daily for 6 months. As expected, oestradiol increased in the control and androgen-alone groups but remained suppressed in the group also receiving letrozole. There was a significant increase in predicted adult height in the letrozole group compared with the androgen-alone group and testis volume was more markedly increased compared with the controls. In another study [59] eight boys with delayed puberty were given testosterone enanthate (1 mg/kg) for 6 months and 2.5 mg of letrozole for 12 months. Oestrogen was suppressed in the letrozole group but virilization occurred and puberty was accelerated. Letrozole given with growth hormone also enabled improvement in height in an adolescent with short stature [60].

It is concluded that androgen combined with an aromatase inhibitor is a beneficial treatment for boys with delayed puberty. The former causes virilization and a growth spurt and the latter prevents epiphyseal closure so allowing a longer duration of bone growth. Moreover, the effect may be magnified because combined therapy produces higher serum testosterone and lower serum oestrogen. There are potential concerns that aromatase inhibitors may have adverse effects on metabolism, including on protein, lipid and bone biochemistry. However, these concerns have been largely allayed. It has been clearly demonstrated that in pubertal boys given androgen and letrozole there are no adverse effects on bone mineral density in the lumber spine and femoral neck or on serum makers of bone resorption and formation [58, 61].

Letrozole actually reduced serum insulin in pubertal boys co-administered androgen, suggesting an improvement in insulin sensitivity. Serum insulin-like growth factor 1 (IGF-1) and IGF-binding protein 3 (IGFBP3) increased in the androgen-treatment group but were unchanged in the letrozole group. The only unfavourable change due to letrozole was a small decrease in high-density lipoprotein (HDL) [62]. Similar findings were reported by Mauras et al. [63, 64], who showed that 0.5 or 1 mg of anastrozole daily caused no change in body composition (body mass index, fat mass or fat-free mass), rates of protein synthesis or degradation, carbohydrate, lipid or protein oxidation, muscle strength, calcium kinetics or bone growth factor concentrations. This contrasted with a marked anti-anabolic effect of the GnRH agonist, leuprolide.

It can be concluded that androgen therapy combined with an aromatase inhibitor has benefit in treatment of delayed puberty and that addition of these drugs is unlikely to compromise metabolism. Further, longer, randomized trials are required to place such therapy in perspective.

There is one report of treatment of a young girl with the McCune–Albright syndrome and gonadotrophin-independent precocious puberty with 1 mg of anastrozole daily [65]. Menstruation stopped and accelerated bone age was arrested and predicted adult height increased by 12 cm. It was concluded that anastrozole has beneficial effects in gonadotrophin-independent precocious puberty but this needs to be confirmed in larger randomized studies.

Gynaecomastia during puberty in boys is not an uncommon event and is a consequence of an imbalance between the stimulatory effects of oestrogen and the inhibitory action of androgens at the breast [66]. Investigation of the effects of aromatase inhibitors in this condition is therefore justified. Two studies have reported the effects of 1 mg of anastrozole on pubertal gynaecomastia in pubertal boys. Riepe et al. [67] described marked reductions in breast size in four of five boys treated, with complete disappearance of glandular tissue in one of them; breast tenderness was resolved by 4 weeks. The longer the duration of gynaecomastia before treatment, the smaller the reduction in breast size that was observed. No adverse effects were recorded.

Plourde et al. [68] described results of a much larger, randomized, double-blind, placebo-controlled study in 80 boys with pubertal gynaecomastia. In the group treated with 1 mg of anastrozole daily the drug was well tolerated but at 6 months there was no difference in the number of patients undergoing a reduction in breast size of greater than 50% between the aromatase-inhibitor and placebo groups. However, in this study gynaecomastia had been present for longer than 1 year in 90% of the patients. It can be concluded that aromatase inhibitors may have some value in pubertal gynaecomastia but that therapy must be initiated soon after its appearance. However, this will also need to be confirmed in larger trials.

The limited effect of aromatase inhibitors in pubertal gynaecomastia is consistent with findings of its limited activity in men with prostate cancer and gynaecomastia due to treatment with the antiandrogen, Casodex, which induces a similar imbalance between oestrogen and androgen [69–71].

### *Thyroid goitre*

Multi-nodular goitre is a common thyroid disease, particularly in women, but is usually asymptomatic. Epidemiological observations and experimental data have implicated oestrogen generated within the thyroid by aromatase as a culprit driver of the disease [72, 73]. Consequently, 32 postmenopausal patients with non-toxic multi-nodular goitre were randomized to receive either 1 mg or anastrozole or placebo daily for 3 months; 28 patients completed the study. There was no significant reduction in goitre size and no significant changes in thyroid function, thyroglobulin, gonadotrophins, sex hormone binding, globulin, oestradiol or testosterone [72, 73]. It can be concluded that aromatase inhibitors have no benefit in multi-nodular thyroid goitre.

### *Toxicity*

Third-generation aromatase inhibitors have been generally well tolerated. However, concern has been expressed about the impact of oestrogen withdrawal on bone integrity, lipids, the vascular system and even on cognitive function. These concerns have been addressed in a number of studies in addition to those above where comments on toxicity have already been made. There is evidence that letrozole may not have deleterious effects on bone in mice [74]. An increase in epiphyseal growth-plate height and proliferation of chondrocytes is observed in letrozole-treated peri-pubertal mice and so it can be concluded that letrozole has the potential to increase linear growth and not cause damage to skeletal integrity [74]. Anastrozole given with androgens and finasteride prevented bone loss following orchidectomy of aged rats [75] but it is unclear what the relative contribution of each drug is in this respect.

The steroidal aromatase inhibitor exemestane has also been shown to prevent bone loss and maintain bone strength in ovariectomized rats [76] but it is unclear whether this is primarily due to aromatase inhibition or to the androgenicity of the compound. However, exemestane also appears to prevent bone loss in premenopausal women [77].

In contrast, a majority of papers suggest that the non-steroidal aromatase inhibitors accelerate bone loss. Anastrozole for at least 6 months caused bone loss by radiometric assessment in postmenopausal women with breast cancer [78–80]. Similarly, anastrozole caused increases in markers of bone resorption and decreases in markers of bone formation in elderly men during 3 weeks of treatment [81, 82]. However, Leder et al. [81] have claimed that anastrozole does not adversely affect bone metabolism in elderly men based on assessment of bone turnover. The reasons for these discrepant results are unclear.

Letrozole has been consistently shown to accelerate bone loss but there was no evidence of increased fracture rates in women with breast cancer after a median of 2.4 years follow-up [84]. However, it is unlikely that fracture rate would increase in such a short period. Similarly, letrozole increases markers of bone turnover in healthy postmenopausal women [85].

Even if further long-term studies do indicate that the non-steroidal aromatase inhibitors cause enhanced bone loss, co-administration of calcium, vitamin D or bis-phosphonates should overcome any issues and can certainly be justified in the treatment of malignant disease.

The three large randomized adjuvant therapy trials with anastrozole [86], letrozole [84] and exemestane [87] allow detailed analysis of the cardiovascular effects: these have been reviewed by Howell and Cuzick [88], who emphasized that caution must be exercised in interpretation as the studies are still in progress.

In the exemestane trial there was a significantly greater number of coronary deaths in the aromatase-inhibitor group than in patients randomized to tamoxifen [87]. It is unclear whether this is the result of a genuine adverse impact of exemestane or a protective effect of tamoxifen. More coronary events (chest pain, angina and myocardial infarcts) were also seen in the ATAC trial [86] compared with tamoxifen but this was non-significant. Similarly, there were more coronary events and deaths in the letrozole arm than in the tamoxifen group but again this was non-significant [84].

The effects of aromatase inhibitors on lipids have been mainly studied in these breast cancer trials. Anastrozole has not been associated with major effects on lipid profiles. However, there is a report of increased HDL and decreased triglycerides [89]. Results from letrozole studies have been conflicting. In a study of 20 patients given letrozole significant increases in cholesterol and low-density lipoprotein-cholesterol were observed [90]. In contrast Harper-Wynne et al. [91] and Goss et al. [84] reported no effect on lipid profiles. Similarly, two studies with exemestane in patients with advanced breast cancer have yielded conflicting results. Nine weeks of treatment resulted in a significant reduction in cholesterol and total triglycerides but also an unfavourable reduction in HDL [92], whereas there were no changes in cholesterol, HDL, apolipoproteins A1 or B, or lipoprotein (a) in a more recent study [93, 94].

It is essential that long-term studies with all three third-generation aromatase inhibitors are carried out in previously untreated patients to determine the real impact on coronary disease and serum lipids so that these conflicting data can be resolved. The only report of an effect of an aromatase inhibitor on cerebrovascular disease concerns anastrozole. In the ATAC trial [86] there were significantly fewer cerebrovascular accidents in patients given anastrozole than those randomized to tamoxifen. Again, it is unclear whether this is due to a protective effect of anastrozole or an enhanced event rate in the tamoxifen group. In both the ATAC trial [86] and the exemestane study [87] there were fewer thromboembolic events in the aromatase inhibitor arms than in the tamoxifen groups.

There are concerns that reductions in oestrogen will impact negatively on cognitive function. None of the major studies has reported on cognitive function but there is a sub-protocol on cognitive function in the International Breast Cancer Intervention Study II that compares anastrozole with placebo. There is a single small study in elderly men given testosterone with or without anastrozole [95]. Interestingly, improvements in verbal memory but not spatial ability were seen with testosterone alone whereas addition of anastrozole prevented the improvement in verbal memory but caused an increase in spatial ability. This suggests that oestrogen may contribute positively to verbal memory but have adverse effects on spatial ability. These observations will need to be confirmed in more extensive studies.

There are clearly many discussions in the endocrine and cancer community about the safety of aromatase inhibitors, but overall there is little evidence that should cause real concern. In malignant disease any risk is far outweighed by the benefits and aromatase inhibitors stand comparison well with the best of other therapeutic regimes. In the benign diseases aromatase inhibitors still appear to have a favourable risk/benefit ratio, where they are effective.

## Conclusions

The following conclusions relate largely to the non-steroidal aromatase inhibitors, although it would be wrong to conclude that they are identical in their actions and side effects. Exemestane, as a steroid, may have other properties that might make it either more or less suitable in the diseases listed below.

Aromatase inhibitors have no major effects in malignant disease apart from breast cancer. There is some modest activity in ovarian and endometrial cancer but it is clear that only patients who have tumours that are ER positive are likely to responsive. Patients with ovarian cancer who have not been heavily pretreated and those with endometrial cancer who have not received progestin therapy may also be more likely to respond. In contrast, endometrial stromal sarcoma, a relatively uncommon tumour, seems to be more amenable to therapy with aromatase inhibitors. There is no benefit of aromatase inhibitors in prostatic and liver cancer.

Aromatase inhibitors do have a role to play in female infertility, where in most studies their efficacy compares favourably with clomiphene and tamoxifen. Aromatase inhibitors appear to have the advantage that they can induce ovulations yet allow full development of the endometrium; this is not the case with the antioestrogens that limit endometrial height. Moreover, fewer ampoules of expensive FSH preparations are required if aromatase inhibitors are co-administered.

There is some evidence that aromatase inhibitors can be effective in endometriosis, particularly in women with minimal or no ovarian function. In premenopausal women there are good responses if the aromatase inhibitors are given with a GnRH agonist, like Zoladex, but the comparative value of the two drugs in this condition is unclear. It is noteworthy that the combination seems to be associated with lower and slower relapse rates after drug withdrawal. The precise scheduling of the drugs to cause optimal responses has yet to be determined. There are very limited data in treatment of fibromatosis so this is worthy of further study.

In men, aromatase inhibitors cause an increase in circulating LH and testosterone and a reduction in oestrogen. Thus, such therapy may be of value in some infertile men and in ageing men with waning testis function, but the physiological significance of the endocrine changes is yet to be determined.

Aromatase inhibitors are effective in adolescents with delayed puberty when given with androgens. Virilization ensues but the aromatase inhibitor prevents epiphyseal closure and so increases the predicted adult height. There is very limited evidence for a role of aromatase inhibitors in precocious puberty but good responses were seen in a girl with McCune–Albright syndrome. There are also some responses in pubertal gynaecomastia but it is evident that aromatase inhibitors have to be given early in the condition to produce optimal effects. Aromatase inhibitors have no value in multi-nodular thyroid goitre.

There should be no doubt that aromatase inhibitors are well tolerated. However, the main tolerance issues that are frequently discussed relate to increased bone resorption and possible cardiovascular and cognitive function effects. The majority of studies show that aromatase inhibitors do increase bone turnover and reduce bone mineral density but have not been shown to increase fracture rate. Moreover, the effects are not as marked as with GnRH agonists. The risk/benefit ratio in malignant disease is clearly very favourable and should be of no concern in benign diseases where intermittent therapy is common. In any case, calcium, vitamin D and bis-phosphonates may all prevent or reverse any bone loss that does occur. The impact of aromatase inhibitors on coronary events and lipid profiles is difficult to assess because the comparator is usually tamoxifen, which has effects of its own, but it is clear that there are no major negative impacts on lipid metabolism in breast cancer patients. There are very limited data on effects on cognitive function but there is nothing to suggest that this is seriously impaired.

## References


1 Furr BJA, Jordan VC (1984) The pharmacology and clinical uses of tamoxifen. *Pharmac Therap* 25: 127–205

2 Bowman A, Gabra H, Langdon SP, Lessells A et al. (2002) CA125 response is associated with oestrogen receptor expression in a Phase II trial of letrozole in ovarian cancer: identification of an endocrine-sensitive subgroup. *Clin Cancer Res* 8: 2233–2239

3 Papadimitriou CA, Markaki S, Siapkarasw J, Vlachos G et al. (2004) Hormonal therapy with letrozole for relapsed epithelial ovarian cancer. *Oncology* 66: 112–117

4 Krasner CN, Debernado RL, Findley M, Penson R et al. (2005) Phase II trial of ansastrazole in Combination with Gefitinib in Women with Asymtomatic Mullerian Cancer. *Proc Am Soc Clin Oncol* Abstract 5063

5 Sasano H, Sata S, Ito K, Yajima A et al. (1999) Effects of aromatase inhibitors on the pathobiology of the human breast, endometrial and ovarian carcinoma. *Endocr Relat Cancer* 6: 197–204

6 Morsi H, Leers M, Nap M, Bjorklund V et al. (2000) Apoptosis and anti-apoptosis in oestrogen-receptor negative endometrial cancer cells in response to anastrozole, 4-hydroxytamoxifen and medroxyprogesterone acetate. *Eur J Cancer* 36 (Suppl 4): 112–113

7 Rose PG, Brunetto VL, Vanle L, Bell J et al. (2000) A phase II trial of anastrozole in advanced recurrent or persistent edometrial carcinoma: a Gynecologic Oncology Group Study. *Gynecol Oncol* 78: 212–216

8 Berstein L, Maximov S, Gersgfeld E, Meshkova I et al. (2002) Neoadjuvant therapy of endometrial cancer with the aromatase inhibitor letrozole: endocrine and clinical effects. *Eur J Obstet Gynecol Reprod Biol* 105: 161–165

9 Ma BBY, Oza A, Eisenhauser E, Stanimir G et al. (2004) The activity of letrozole in patients with advanced or recurrent endometrial cancer and correlation with biological markers – a study of the National Cancer Institute of Canada Clinical Trials Group. *Int J Gynecol Cancer* 14: 650–658

10 Reich O, Regauger S, Urdl W, Lahousen M et al. (2000) Expression of oestrogen and progesterone receptors in low-grade endometrial stromal sarcomas. *Br J Cancer* 82: 1030–1034

11 Reich O, Regauer S (2004) Aromatase expression in low-grade endometrial stromal sarcomas. A immunohistochemical study. Mod Pathol 17: 104–108

12 Maluf FC, Sabbatini MD, Schwartz L, Xia J et al. (2001) Endometrial stromal sarcoma: objective response to letrozole. *Gynecol Oncol* 82: 384–388

13 Leunen M, Breugelmans M, De Sutter PH, Bourgain C et al. (2004) Low grade endometrial stromal sarcoma treated with the aromatase inhibitor letrozole: case report. *Gynecol Oncol* 95: 769–771

14 Spano JP, Soria JC, Kambouchner M, Piperno-Neuman et al. (2003) Long term survival of patients given hormonal therapy for endometrial stromal sarcoma. *Med Oncol* 94: 87–94

15 Reich O, Regauer S (2005) Hormonal treatment with aromatase inhibitors for patients with endometrial sarcoma. *Gynecol Oncol* 98: 173–174

16 Santen RJ, Petroni GR, Fisch MJ, Myers CE et al. (2001) Use of the aromatase inhibitor anastrozole in the treatment of patients with advanced prostate carcinoma. *Cancer* 92: 2095–2101

17 Smith MR, Kaufman D, George D, Oh WK et al. (2002) Selective aromatase inhibition for patients with androgen-independent prostate carcinoma: a Phase II study of letrozole. Cancer 95: 1864–1868

18 Bonomo M, Mingrone W, Brauchli P, Hering F et al. (2003) Exemestane seems to stimulate growth in men with prostate carcinoma. *Eur J Cancer* 39: 2111–2112

19 Grosh WW, Mueller S, Alexander BF, Caldwell SH et al. (1999) Anastrozole (A) treatment of patients (pts) with advanced unresectable hepatocellular carcinoma (HCC). Proc ASCO 18: 286a Abstract 1096

20 Mitwally MFM, Casper RF (2000) The aromatase inhibitor, Letrozole: a promising alternative for clomiphene citrate for induction of ovulation. *Fertil Steril* 74: S35–S37
21 Mitwally MFM, Casper RF (2001) Use of an aromatase inhibitor for induction of ovulation in patients with an inadequate response to clomiphene citrate. *Fertil Steril* 75: 305–309
22 Al-Omari WR, Sulaiman WR, Al-Hadithi N (2003) Comparison of two aromatase inhibitors in women with clomiphene-resistant polycystic ovary syndrome. Gynecol Obstet 85: 289–291
23 Park WI, Shin SY, Yang JS, Lee JY (2004) Clinical outcomes of ovulation induction with aromatase inhibitor: Prospective randomised comparison with clomiphene citrate. *J Obstet Gynaecol Res* 30: 262 Abstract 8–19
24 Elnashar AM (2002) Current uses of aromatase inhibitors in gynecology. *Middle East Fertil Soc* J 7: 173–179
25 Fisher SA, Reid MD, Van Vugt DA, Casper RF (2002) A randomised double-blind comparison of the effects of clomiphene citrate and the aromatase inhibitor letrozole on ovulatory function in normal woman. *Fertil Steril* 78: 280–285
26 Sammour A, Biljan MM, Tan SL, Tulandi T (2001) Prospective randomised trial comparing the effects of letrozole and clomiphene citrate on follicular development, endometrial thickness and pregnancy rate in patients undergoing superovulation prior to intrauterine insemination. *Fertil Steril* 76: S110–S112
27 Mousavi-Fatemi H, Kolibiankis E, Tournaye H, Camus M et al. (2003) Clomiphene citrate versus letrozole for ovarian stimulation: a pilot study. *Reprod Bio Med Online* 7: 543–546
28 El Helw B, El Sadek M, Matar H, Fouad S et al. (2002) Single dose Letrozole versus CC for superovulation prior to intrauterine insemination: a prospective randomised study. *Hum Reprod* 17: Abstract 73
29 Mitwally MFM, Casper RF (2002) Aromatase inhibition improves ovarian response to follicle-stimulating hormone in poor responders. *Fertil Steril* 77: 776–780
30 Mitwally MFM, Casper RF (2003) Aromatase inhibition reduced gonadotrophin dose required for controlled ovarian stimulation in women with unexplained infertility. *Hum Reprod* 18: 1588–1597
31 Mitwally MFM, Casper RF (2004) Aromatase inhibition reduces the dose of gonadotropin required for controlled ovarian hyperstimulation. *J Soc Gynaecol Invest* 11: 406–415
32 Tsirigotis M, Prapas G, Pistofidis G, Tika M et al. (2002) Experience with the use of letrozole in patients with a history of poor ovarian response in previous assisted reproductive cycles. *Hum Reprod* 17: Abstract 34
33 Healey S, Lin Tan S, Tulandi T, Biljan MM (2003) Effects of letrozole on superovulation with gonadotropins in women undergoing intrauterine insemination. *Fertil Steril* 80: 1325–1329
34 Casper RF (2003) Letrozole: ovulation or superovulation? *Fertil Steril* 80: 1335–1337
35 Noble LS, Simpson ER, Johns A, Bulun SE (1996) Aromatase expression in endometriosis. *J Clin Endocr Metab* 81: 174–179
36 Kitawaki J, Noguchi T, Amatsu T, Maeda K et al. (1997) Expression of aromatse cytochrome P450 protein and messenger ribonucleic acid in human endometriotic and adenomyotic tissues but in normal endometrium. *Biol Reprod* 57: 514–519
37 Bulun SE, Yang S, Fang Z, Gurates B et al. (2001) Role of of aromatase in endometrial disease. *J Steroid Biochem Mol Biol* 79: 19–25
38 Takayama K, Zeitoun K, Gunby RT, Sasano H et al. (1998) Treatment of severe postmenopausal endometriosis with an aromatase inhibitor. *Fertil Steril* 69: 709–713
39 Zeitoun K, Takayama K, Gunby RT, Carr BR et al. (1998) Treatment of severe endometriosis with an aromatase inhibitor: A novel indication and its molecular basis. *J Soc Gynecol Invest* 5 (Suppl): 96a
40 Razzi S, Fava A, Sartini A, De Simone S et al. (2004) Treatment of severe recurrent endometriosis with an aromatase inhibitor in a young ovariectomized woman. *Br J Obstet Gynaecol* 111: 182–184
41 Scarpellini F, Sbracia M (1998) Aromatase Inhibitors in the treatment of severe endometriosis. *Fertil Steril* 70 (Suppl 1): Abstract 009
42 Scarpellini F, Sbracia M (2000) Aromatase inhibitors in the treatment of severe endometriosis. *Proc ESHRE* Abstract O-097
43 Scarpellini F, Sbracia M (2000) Aromatase inhibitors in the treatment of low responder women with severe endometriosis. *Fertil Steril* 72 (Suppl): Abstract P-384
44 Ailawadi RK, Jobanputra S, Kataria M, Gurates B et al. (2004) Treatment of endometriosis and chronic pelvic pain with letrozole and norethindrone acetate: a pilot study. *Fertil Steril* 81: 290–296

45 Soysal S, Soysal ME, Ozer S, Gul N et al. (2004) The effects of post-surgical administration of goserelin plus anastrozole compared to goserelin alone in patients with severe endometriosis: a prospective randomised trial. *Hum Reprod* 19: 160–167
46 Shippen ER, West WJ (2004) Successful treatment of severe endometriosis in two premenopausal women with an aromatase inhibitor. *Fertil Steril* 81: 1395–1398
47 Klemi P, Alanen K, Hietanen S, Grenman S et al. (2003) Response of estrogen receptor-positive intraabdominal fibromatosis to aromatase inhibitor therapy. *Obstet Gynecol* 102: 1155–1158
48 Turner KJ, Morley M, Atanassova N, Swanston ID et al. (2000) Effect of chronic administration of an aromatase inhibitor to adult male rats on pituitary and testicular function and fertility. *J Endocr* 164: 225–238
49 Raman JD, Schlegel PN (2002) Aromatase inhibitors for male infertility. *J Urol* 167: 624–629
50 Holbrook JM, Cohen PG (2003) Aromatase inhibition for the treatment of idiopathic hypogonadotropic hypogonadism in men with premature ejaculation. *South Med J* 96: 544–547
51 Leder BZ, Rohrer JL, Rubin SD, Gallo J et al. (2004) Effects of aromatase inhibition in elderly men with low or borderline-low serum testosterone levels. *J Clin Endocr Metab* 89: 1174–1180
52 Finkelstein JS, Neer RM, Beverley MD, Biller MK et al. (1992) Osteopenia in men with a history of delayed puberty. *N Engl J Med* 326: 600–604
53 Uruena M, Pantsiotou S, Preece MA, Stanhope R (1992) Is testosterone therapy for boys with constitutional delay of growth and puberty associated with impaired final height and suppression of the hypothalamo-pituitary-gonadal axis? *Eur J Pediatr* 151: 15–18
54 Stanhope R, Brook CGD (1985) Oxandrolone in low dose for constitutional delay of growth and puberty in boys. *Arch Dis Child* 60: 379–381
55 Carani C, Qin K, Simoni M, Faustini-Fustini M et al. (1997) ER: Effect of testosterone and estradiol in a man with aromatase deficiency. *N Engl J Med* 337: 91–95
56 Bilezikian JP, Morishima A, Bell J, Grumbach MM (1998) Increased bone mass as a result of oestrogen therapy in a man with aromatase deficiency. *N Engl J Med* 339: 599–603
57 Wickman S, Sipila I, Ankarberg-Lindgren C, Norjavaara E et al. (2001) A specific aromatase inhibitor and potential increase in adult height in boys with delayed puberty: a randomised controlled trial. *Lancet* 357: 1743–1748
58 Dunkel L, Wickman S (2002) Novel treatment of delayed male puberty with aromatase inhibitors. *Horm Res* 57 (Suppl 2): 44–52
59 Raivio T, Dunkel L, Wickman S, Janne OA (2004) Serum androgen bioactivity in adolescence: a longitudinal study of boys with constitutional delay of puberty. *J Clin Endocr Metab* 89: 1188–1192
60 Zhou P, Shah B, Prasad K, David R (2005) Letrozole significantly improves growth potential in a pubertal boy with growth hormone deficiency. *Pediatrics* 115: E245–E249
61 Wickman S, Kanjantie E, Dunkel L (2003) Effects of suppression of oestrogen action by the P450 aromatase inhibitor letrozole on bone mineral density and bone turnover in pubertal boys. *J Clin Endocr Metab* 88: 3785–3793
62 Wickman S, Saukkonen T, Dunkel L (2002) The role of sex steroids in the regulation of insulin sensitivity and serum lipid concentrations during male puberty: a prospective study with a P450 aromatase inhibitor. *Eur J Endocr* 146: 339–346
63 Mauras N, O'Brien KO, Klein KO, Hayes V (2000) Estrogen suppression in males: metabolic effects. *J Clin Endocr Metab* 85: 2370–2377
64 Mauras N, Hayes V, O'Brien KO (2000) Estrogen treatment and estrogen suppression: metabolic effects in adolescence. *J Paediatr Endocr Metab* 13: 1431–1437
65 Roth C, Freiberg C, Zappel H, Albers N (2002) Effective aromatase inhibition by anastrozole in a patient with gonadotropin-independent precocious puberty in McCune-Albright Syndrome. *J Paediatr Endocr Metab* 15 (Suppl 3): 945–948
66 Braunstein G (1993) Gynecomastia. *N Engl J Med* 328: 490–495
67 Riepe FG, Baus I, Wiest S, Krone N et al. (2004) Treatment of pubertal gynaecomastia with the specific Aromatase Inhibitor Anastrozole. *Horm Res* 62: 113–118
68 Plourde PV, Reiter EO, Jou H-C, Desrochers PE et al. (2004) Safety and efficacy of Anastrozole for the treatment of pubertal gynecomastia: a randomised, double-blind, placebo-controlled trial. *J Clin Endocr Metab* 89: 4428–4433
69 Saltzstein D, Cantwell A, Sieber P, Ross JR et al. (2002) Prophylactic tamoxifen significantly reduced the incidence of bicalutamide-induced gynaecomastia and breast pain. *BJU International* 90 (Suppl 2): 120–121

70 Boccardo F, Rubagotti A, Garofalo L, Di Tonni P et al. (2003) Tamoxifen (T) is more effective than anastrozole (A) in preventing gynaecomastia induced by bicalutamide (B) monotherapy in prostate cancer (pca) patients (pts). *ASCO Proc* 22: 400 Abstract 1608
71 Conti G, Cretarola E, Boccardo F, Battaglia et al. (2004) Tamoxifen is safe and effective in preventing gynecomastia and breast pain induced by bicalutamide monotherapy of prostate cancer and does not alter treatment efficacy. *Eur Urol* (Suppl) 3: 58
72 Knight JS, Satchidanand R, Cummings MH, Jackson A et al. (2004) A double-blind, randomised, controlled trial to evaluate the effect of anastrozole for the treatment of non-toxic multinodular goitre. *Br J Surg* 91 (Suppl 1): 57
73 Knight HS, Yagat R, Yiangou C, Jackson A (2004) A double blind randomised controlled trial to evaluate the effect of anastrozole on non-toxic multinodular goitre. *Clin Sci* 106: 9P
74 Eshet R, Maor G, Ben Ari T, Eliezer MB et al. (2004) The aromatase inhibitor letrozole increases epiphyseal growth plate height and tibial length in peripubertal male mice. *J Endocrinol* 182: 165–172
75 Shih C, Pan MH, Chu P (2000) Effects of estradiol and testosterone/proscar/arimidex on ORX-induced osteopenia in aged rats. *J Bone Min Res* 15: S343
76 Goss PE, Qi S, Josse RG, Pritzker ?? et al. (2004) The steroidal aromatase inhibitor exemestane prevents bone loss in ovariectomized rats. Bone 34: 384–392
77 Goss PE, Thomson T, Banke-Bochita J, Lowery C et al. (2002) A randomised placebo-controlled explorative study to investigate the effect of low oestrogen plasma levels on markers of bone turnover in healthy postmenopausal women during the 12 week treatment with exemestane or letrozole. *Breast Cancer Res Treat* 76 (Suppl 1): S76
78 Baum M, Budzar AU, Cuczick J, Forbes J et al. (2002) Anastrozole alone or in combination with tamoxifen versus tamoxifen alone for adjuvant treatment of postmenopausal women with early breast cancer; first results of the ATAC randomised trial. *Lancet* 359: 2131–2139 (Erratum, *Lancet* 360: 1520)
79 Eastell R, Adams J (2002) Results of the 'Arimidex' (anastrozole, A), tamoxifen (T), alone or in combination (C ) (ATAC) trial: effects on bone mineral density (BMD) and bone turnover. *Ann Oncol* 13 (Suppl 5): 32
80 Wojtacki J, Zielinski KW, Lesniewski-Kmak K, Wiraszka R et al. (2003) Effect of anastrozole therapy on bone: preliminary results of digital radiometrical analysis of clavicle and rib. *Eur J Cancer* (Suppl 1): S120
81 Lai JM, Taxel P, Raisz LG (1998) The effect of aromatase inhibition on bone turnover in older men. *J Am Ger Soc* 46: S79
82 Taxel P, Kennedy DG, Fall PM, Willard AKet al. (2001) The effect of aromatase inhibition on sex steroids, Gonadotropins, and markers of bone turnover in older men. *J Clin Endocr Metab* 86: 2869–2874
83 Leder BZ, Rohrer JL, Rubin SD, Gallo J et al. (2003) Aromatase inhibition does not alter bone turnover, serum OPG or bone density in elderly men with mild hypogonadism. *J Bone Min Res* 18 (Suppl 2): S42
84 Goss PE, Ingle JN, Martino S, Rober NJ et al. (2003) A randomised trial of letrozole in postmenopausal women after five years of tamoxifen therapy for early-stage breast cancer. *N Engl J Med* 349: 1793–1802
85 Harper-Wynne C, Ross G, Sacks N, Dowsett M (2001) A pilot prevention study of the aromatase inhibitor letrozole: effects on breast cell proliferation and bone/lipid indices in healthy postmenopausal women. *Breast Cancer Res Treat* 69: 225
86 ATAC Trialists' Group (2003) Anastrozole alone or in combination with tamoxifen versus tamoxifen alone for adjuvant treatment of postmenopausal women with early-stage breast cancer. *Cancer* 98: 1802–1810
87 Coombes R, Hall E, Gibson LJ, Paridaens R et al. (2004) A randomized trial of exemestane after two or three years of tamoxifen therapy in postmenopausal women with primary breast cancer. *N Engl J Med* 350: 1081–1092
88 Howell A, Cuzick J (2005) Vascular effects of aromatase inhibitors; data from clinical trials. *J Steroid Biochem Mol Biol* 95: 143–149
89 Sawada S, Sato K (2003) Effect of anastrozole and tamoxifen on serum lipid levels in Japanese postmenopausal women with early breast cancer. *Breast Cancer Res Treat* 82 (Suppl 1): S31–S32
90 Elisaf MS, Bairaktari ET, Nicolaides C, Kakaidi B et al. (2001) Effect of letrozole on the lipid profile in postmenopausal women with breast cancer. *Eur J Cancer* 37: 1510–1513

91 Harper-Wynne C, Ross G, Sacks N, Salters J et al. (2002) Effects of the aromatase inhibitor letrozole on normal breast epithelial cell proliferation and metabolic indices in postmenopausal women: a pilot study for breast cancer prevention. *Cancer Epidemiol Biomarkers Prev* 11: 614–621

92 Engen T, Krane J, Johannessen DC, Lonning PE et al. (1995) Plasma changes in breast cancer patients during endocrine therapy – lipid measurements and nuclear magnetic resonance (NMR) spectroscopy. *Breast Cancer Res Treat* 36: 287–297

93 Lohrisch C, Paridaens R, Dirix LY, Beex L et al. (2001) No adverse impact on serum lipids of the irreversible aromatase inactivator Aromasin (Exemestane (E) in 1st line treatment of metastatic breast cancer (MBC): companion study to a European Organisation of Research and Treatment of Cancer. (Breast Group) trial with exemestane. *Proc Am Soc Clin Oncol* 20: 43a

94 Atalay G, Dirix L, Biganzoli L, Beex L et al. (2004) The effect of exemestane on serum lipid profile in postmenopausal women with metastatic breast cancer: a companion study to EORTC trial 1095, "randomised phase II study in first line hormonal treatment for metastatic breast cancer with exemestane or tamoxifen in postmenopausal patients". *Ann Oncol* 15: 211–217

95 Cherrier MM, Asthana S, Plymate S, Baker LD et al. (2000) Cognitive effects from exogenous manipulation of testosterone and estradiol in older men. *Soc Neurosci* 26: 13

# Index

# The MDT-Series
# Milestones in Drug Therapy

The discovery of drugs is still an unpredictable process. Breakthroughs are often the result of a combination of factors, including serendipidity, rational strategies and a few individuals with novel ideas. *Milestones in Drug Therapy* highlights new therapeutic developments that have provided significant steps forward in the fight against disease. Each book deals with an individual drug or drug class that has altered the approach to therapy. Emphasis is placed on the scientific background to the discoveries and the development of the therapy, with an overview of the current state of knowledge provided by experts in the field, revealing also the personal stories behind these milestone developments. The series is aimed at a broad readership, covering biotechnology, biochemistry, pharmacology and clinical therapy.

**Forthcoming titles**

*Tamoxifen and Beyond,* V.C. Jordan (Author), 2006
*Pharmacotherapy of Obesity*, J.P.H. Wilding (Editor), 2006
*TNF-alpha Inhibitors,* J.M. Weinberg, R. Buchholz (Editors), 2006
*Drugs affecting Growth of Tumours*, H.M. Pinedo, C. Smorenburg (Editors), 2006

**Published volumes**

*Cannabinoids as Therapeutics*, R. Mechoulam (Editor), 2005
*St. John`s Wort and its Active Principles in Anxiety and Depression*, W.E. Müller (Editor), 2005
*Drugs for Relapse Prevention of Alcoholism*, R. Spanagel, K. Mann (Editors), 2005
*COX-2 Inhibitors*, M. Pairet, J. Van Ryn (Editors), 2004
*Calcium Channel Blockers*, T. Godfraind (Author), 2004
*Sildenafil*, U. Dunzendorfer (Editor), 2004
*Hepatitis Prevention and Treatment*, J. Colacino, B.A. Heinz (Editors), 2004
*Combination Therapy of AIDS*, E. De Clercq, A.M. Vandamme (Editors), 2004
*Cognitive Enhancing Drugs*, J. Buccafusco (Editor), 2004
*Fluoroquinolone Antibiotics*, A.R. Ronald, D. Low (Editors), 2003
*Erythropoietins and Erythropoiesis*, G. Molineux, M. Foote, S. Elliott (Editors), 2003
*Macrolide Antibiotics*, W. Schönfeld, H. Kirst (Editors), 2002

*HMG CoA Reduktase Inhibitors*, G. Schmitz, M. Torzewski (Editors), 2002
*Antidepressants*, B.E. Leonard (Editor), 2001
*Recombinant Protein Drugs*, P. Buckel (Editor), 2001
*Glucocorticoids,* N. Goulding, R.J. Flower (Editors), 2001
*Modern Immunosuppressives*, H.-J. Schuurman (Editor), 2001
*ACE Inhibitors*, P. D'Orleans-Juste, G. Plante (Editors), 2001
*Atypical Antipsychotics*, A.R. Cools, B.A. Ellenbroek (Editors), 2000
*Methotrexate*, B.N. Cronstein, J.R. Bertino (Editors), 2000
*Anxiolytics*, M. Briley, D. Nutt (Editors), 2000
*Proton Pump Inhibitors*, L. Olbe (Editor), 1999
*Valproate*, W. Löscher (Editor), 1999